Machine Tools

This book introduces the applications of Industry 4.0 in machine tools through an overview of the latest available digital technologies. It focuses on digital twining, communication between industrial controls, motion, and input/output devices, along with sustainability in SMEs.

Machine Tools: An Industry 4.0 Perspective focuses on the digital twining of machine tools, which improves the life of the machines and provides a method of operating a factory during times of complete lockdown resulting from various conditions. It presents an overview of the communication between industrial controls, motion and input/output devices through standardized digital interfaces such as SERCOS and USB. The book goes on to discuss industrial cybersecurity systems applicable to discrete manufacturing, which include cyberattacks and human errors and address the security aspects related to software, hardware and data. The book also explores the application of big data for different stages of production and illustrates the uses such as predictive maintenance, product quality, product life cycle management (PLM) and more.

This book is an ideal reference for undergraduate, graduate, and postgraduate students of industrial, mechanical and mechatronics engineering, along with professionals and general readers.

Computers in Engineering Design and Manufacturing

Series Editor:
Wasim Ahmed Khan
GIK Institute of Engineering Sciences and Technology, Topi, Pakistan

Functional Reverse Engineering of Machine Tools
*Edited by Wasim Ahmed Khan, Ghulam Abbas,
Khalid Rahman, Ghulam Hussain, Cedric Aimal Edwin*

**Functional Reverse Engineering of Strategic
and Non-Strategic Machine Tools**
*Edited by Wasim Ahmed Khan, Khalid Rahman,
Ghulam Hussain, and Ghulam Abbas*

Machine Tools
An Industry 4.0 Perspective
*Edited by Wasim Ahmed Khan, Khalid Rahman, Ghulam Hussain,
Ghulam Abbas, and Wang Xiaoping*

For more information on this series, please visit: www.routledge.com/Computers-in-Engineering-Design-and-Manufacturing/book-series/CRCCOMENGDES

Machine Tools

An Industry 4.0 Perspective

Edited by Wasim Ahmed Khan, Khalid Rahman,
Ghulam Hussain, Ghulam Abbas
and Wang Xiaoping

CRC Press
Taylor & Francis Group
Boca Raton London New York

CRC Press is an imprint of the
Taylor & Francis Group, an **informa** business

First edition published 2023
by CRC Press
6000 Broken Sound Parkway NW, Suite 300, Boca Raton, FL 33487–2742

and by CRC Press
4 Park Square, Milton Park, Abingdon, Oxon, OX14 4RN

CRC Press is an imprint of Taylor & Francis Group, LLC

© 2023 selection and editorial matter, Wasim Ahmed Khan, Khalid Rahman, Ghulam Hussain, Ghulam Abbas, and Wang Xiaoping; individual chapters, the contributors

Library of Congress Cataloging-in-Publication Data
Names: International Workshop on Functional Reverse Engineering of Machine Tools (3rd : 2022 : Ghulam Ishaq Khan Institute of Engineering Sciences and Technology (Pakistan)) | Khan, Wasim A., editor. | Ghulam Ishaq Khan Institute of Engineering Sciences and Technology (Pakistan), host institution.
Title: Machine tools : an industry 4.0 perspective / edited by Wasim A. Khan, Khalid Rahman, Ghulam Hussain, Ghulam Abbas, Wang Xiaoping
Description: First edition. | Boca Raton, FL : CRC Press, 2023. | Series: Computers in engineering design & manufacturing | "The present work is the output of the Third International Workshop on Functional Reverse Engineering of Machine Tools. The workshop, for which Professor Khan kindly asked me to be the Honorary Chair, was a successful hybrid meeting held both online and in-person at Ghulam Ishaq Khan (GIK) Institute of Engineering Sciences and Technology"—Preface. | Includes bibliographical references and index.
Identifiers: LCCN 2022036268 (print) | LCCN 2022036269 (ebook) | ISBN 9781032116693 (hbk) | ISBN 9781032116709 (pbk) | ISBN 9781003220985 (ebk)
Subjects: LCSH: Machine-tools—Congresses. | Machining—Data processing—Congresses. | Industry 4.0—Congresses.
Classification: LCC TJ1185 .I578 2022 (print) | LCC TJ1185 (ebook) | DDC 621.9/02—dc23/eng/20221012
LC record available at https://lccn.loc.gov/2022036268
LC ebook record available at https://lccn.loc.gov/2022036269

ISBN: 978-1-032-11669-3 (hbk)
ISBN: 978-1-032-11670-9 (pbk)
ISBN: 978-1-003-22098-5 (ebk)

DOI: 10.1201/9781003220985

Typeset in Times
by Apex CoVantage, LLC

Mr. Jehangir Bashar
For encouraging us, always

Contents

Foreword

This book is the third in a series edited by Professor Wasim A. Khan and his colleagues focusing on functional reverse engineering of machine tools. I have known the lead editor Professor Khan for over ten years, mainly through correspondence. He is no stranger to machine tools and advanced manufacturing, having authored many publications on these subjects, including a popular monograph on virtual manufacturing for my Springer Series in Advanced Manufacturing.

The present work is the output of the Third International Workshop on Functional Reverse Engineering of Machine Tools. The workshop, for which Professor Khan kindly asked me to be the honorary chair, was a successful hybrid meeting held both online and in-person at Ghulam Ishaq Khan (GIK) Institute of Engineering Sciences and Technology. The range of topics covered was broader than the title of the event implies and included the internet of things, robotics, artificial intelligence, augmented reality and additive manufacturing, all key enabling Industry 4.0 technologies.

I understand that the initial motivation for functional reverse engineering of CNC machine tools might be to help the so-called N11 countries—the next emerging eleven economies—to learn how to design and build advanced manufacturing equipment. However, in isolation, 'reverse engineering' which is a euphemism for 'copying' would at best just allow technological catching up. It is through innovation afforded by the creative and appropriate use of advanced tools such as those associated with Industry 4.0 that leapfrogging could happen. This book should thus be of particular interest to readers from the N11 countries wishing to study some of the latest developments in the field that might enable their companies to leapfrog and compete against those in the industrialized nations.

D. T. Pham, OBE FREng FLSW FSME PhD DEng CEng FIET FIMechE
Chance Professor of Engineering
University of Birmingham
England

Preface

After having two books related to functional reverse engineering in the series, the third volume aims at implementation of fourth Industrial Revolution technologies. *Machine Tools: An Industry 4.0 Perspective* refers to topics from engineering design to modern control. It investigates the theoretical approaches to implementation through standards—both legacy and Industry 4.0. The title describes the methods of prototyping of machine tools and then discusses ways to convert the prototypes to industrial scale machine tool.

Like two earlier books on the topic, this book also describes matters in self-contained modular chapters covering functional reverse engineering of a strategic or non-strategic machine tool's part, structure, assembly, mechanism or whole machine. The value addition made here is introduction of one or multiple fourth industrial revolution technologies embedded in the system engineering design and manufacturing of these machine tool.

The book is a suitable reading for senior year undergraduate students, graduate students and postgraduate students as source of information, solution as a case study and implementation as a prototype respectively.

The next book covers integration of machine tools to form a cell or a manufacturing system in a heterogeneous legacy and modern technological environment.

Editors
Topi, June 2022

About the Editors

Professor Wasim A. Khan has researched and developed industrial scale subtractive and additive manufacturing machine tools, measurement, and testing machines as well as digital controllers for this equipment. He is instrumental in policies and framework development related to manufacturing of discrete products in small, medium, and large-scale manufacturing enterprises at national level. Dr. Wasim is a life member of Pakistan Engineering Council, a chartered engineer of Engineering Council, UK and a fellow of Institution of Mechanical Engineers, UK. He is also a senior member of IEEE, USA. He is currently acting as professor and pro-rector (academics) at the GIK Institute of Engineering Sciences and Technology, Pakistan.

Professor Khalid Rahman is an associate professor in the faculty of mechanical engineering at Ghulam Ishaq Khan Institute of Engineering Sciences and Technology, where he has been a faculty member since 2012. He received his BS in mechanical engineering from Ghulam Ishaq Khan Institute of Engineering Sciences and Technology, MS and PhD degree from Jeju National University, South Korea in 2012. He also has industrial experience of seven years in design and manufacturing. His research interests include printing technologies, and he is currently working on direct ink write and electrohydrodynamic inkjet printing for fabrication of electronics devices and sensors and applications.

Professor Ghulam Hussain is a professor of mechanical engineering at the University of Bahrain, Bahrain. He has a rich experience of both academia and industry. He is an expert of manufacturing technologies with expertise in plasticity, sustainability, and advanced processes including 3D printing, incremental forming, friction welding and hybridization. He is the author of over one hundred top-notch journal articles and an editor of several books. He is recognized among the top ten leading researcher at the international as well as national level. He has been ranked as a world's top 2% scientist by Stanford University, USA. He is the winner of many research and teaching awards/prizes. He is actively involved with several reputed international universities as a foreign expert and external researcher.

Professor Ghulam Abbas received the BS degree in computer science from University of Peshawar, Pakistan, in 2003, and the MS degree in distributed systems and the PhD degree in computer networks from the University of Liverpool, UK, in 2005 and 2010, respectively. From 2006 to 2010, he was research associate with Liverpool Hope University, UK, where he was associated with the Intelligent & Distributed Systems Laboratory. Since 2011, he has been with the Faculty of Computer Sciences & Engineering, GIK Institute of Engineering Sciences and Technology, Pakistan. He is currently working as associate professor and director at Huawei ICT Academy. Dr. Abbas is a co-founding member of the Telecommunications and Networking Research Center at GIK Institute. He is a fellow of the Institute of

Science & Technology, U.K., a fellow of the British Computer Society, and a senior member of the IEEE. His research interests include computer networks and wireless and mobile communications.

Professor Wang Xiaoping is a professor at college of Electrical and Mechanical Engineering, Nanjing University of Aeronautics and Astronautics, a senior member of Chinese Mechanical Engineering Association, a senior member of Aeronautical Society, a guest editor of international journal IJCAT, an editorial board member of American Journal of Software Engineering and Applications, and an evaluation expert of postgraduate dissertation of Ministry of Education. His main research areas include digital design and manufacturing, and intelligent manufacturing. The current research interests are reverse engineering, aerial measurement, geodetic model reconstruction (3D reconstruction); composite material fiber laying process/robot road planning; NC machining tool rail planning/material reduction and material increase manufacturing path optimization; CAD software system integration and algorithm; mathematical modeling; Other theoretical and technical issues related to mechanical design and manufacturing. He has executed several research projects funded by national agencies. He is also winner of three National Defense Science and Technology Progress Awards, one Provincial Award and one Municipal Award. He has published more than eighty research articles in renowned journals and has two patents to his credit.

Contributors

Ghulam Abbas
Faculty of Computer Science and
  Engineering
GIK Institute of Engineering Sciences
  and Technology
Topi, KP, Pakistan

Syed Waqar Ahmed
Department of Mechanical Engineering
Universiti Teknologi PETRONAS
Perak, Malaysia

Daniyal Akram
Faculty of Mechanical Engineering
GIK Institute of Engineering Sciences
  and Technology
Topi, KP, Pakistan

Yousaf Ali
Faculty of Mechanical Engineering
GIK Institute of Engineering Sciences
  and Technology
Topi, KP, Pakistan

Khurram Altaf
Department of Mechanical Engineering
Universiti Teknologi PETRONAS
Perak, Malaysia

Ali Alvi
Faculty of Mechanical Engineering
GIK Institute of Engineering Sciences
  and Technology
Topi, KP, Pakistan

Salman Amin
Faculty of Mechanical Engineering
GIK Institute of Engineering Sciences
  and Technology
Topi, KP, Pakistan

Arsalan Arif
Faculty of Mechanical Engineering
GIK Institute of Engineering Sciences
  and Technology
Topi, KP, Pakistan

Cuneyt Boz
Mechanical Engineering Program
Middle East Technical University—
  Northern Cyprus Campus
Kalkanli, Guzelyurt, Mersin,
  Turkey

Johannes Buhl
Fachgebiet Hybride Fertigung
Fakultät Maschinenbau, Elektro- und
  Energiesysteme
Brandenburgische Technische
  Universität Cottbus—Senftenberg
Cottbus, Germany

Zhengyin Chen
College of Mechanical and Electrical
  Engineering
Nanjing University of Aeronautics and
  Astronautics
Nanjing, China

B. S. Chowdhry
Mehran University of Engineering and
  Technology
Jamshoro, Pakistan

Cedric Aimal Edwin
Department of Management Sciences
  CECOS University of IT and
  Emerging Sciences
Peshawar, Pakistan

Volkan Esat
Mechanical Engineering Program
Middle East Technical University—
    Northern Cyprus Campus
Kalkanli, Guzelyurt, Mersin, Turkey

Muhammad U. Farooq
Department of Mechanical Engineering
Khwaja Fareed University of
    Engineering & Information
    Technology
Rahim Yar Khan, Pakistan

S. S. Farooq
Department of Mechanical Engineering
Khwaja Fareed University of Engineering
    & Information Technology
Rahim Yar Khan, Pakistan

Ghulam Hussain
Faculty of Mechanical Engineering
GIK Institute of Engineering Sciences
    and Technology
Topi, KP, Pakistan
and
Mechanical Engineering Department
College of Engineering
University of Bahrain
Isa Town, Kingdom of Bahrain

Sayed Qaisar Hussain
Faculty of Mechanical Engineering
GIK Institute of Engineering Sciences
    and Technology
Topi, KP, Pakistan

Tanweer Hussain
Mehran University of Engineering and
    Technology
Jamshoro, Pakistan

Ulfat Hussain
Faculty of Mechanical Engineering
GIK Institute of Engineering Sciences
    and Technology
Topi, KP, Pakistan

Abid Imran
Faculty of Mechanical Engineering
GIK Institute of Engineering Sciences
    and Technology
Topi, KP, Pakistan

Rameez Israr
Fachgebiet Hybride Fertigung, Fakultät
    Maschinenbau
Elektro- und Energiesysteme,
    Brandenburgische Technische
    Universität Cottbus—Senftenberg
Cottbus, Germany

Muhammad U. Kaimkhani
Faculty of Mechanical Engineering
GIK Institute of Engineering Sciences
    and Technology
Topi, KP, Pakistan

Zareena Kausar
Department of Mechatronics
    Engineering
Air University
Islamabad, Pakistan

Ghias M. Khan
Department of Mechanical
    Engineering
Khwaja Fareed University of
    Engineering & Information
    Technology
Rahim Yar Khan, Pakistan

Ijlal Ullah Khan
Faculty of Mechanical Engineering
GIK Institute of Engineering Sciences
    and Technology
Topi, KP, Pakistan

Muhammad S. Khan
Faculty of Mechanical Engineering
GIK Institute of Engineering Sciences
    and Technology
Topi, KP, Pakistan

Wasim A. Khan
Faculty of Mechanical Engineering
GIK Institute of Engineering Sciences
    and Technology
Topi, KP, Pakistan

Hao Liu
College of Mechanical and Electrical
    Engineering
Nanjing University of Aeronautics and
    Astronautics
Nanjing, China

Ahmed Murtaza
Faculty of Mechanical Engineering
GIK Institute of Engineering Sciences
    and Technology
Topi, KP, Pakistan

Ali Nasir
Assistant Professor
Control and Instrumentation
    Department Research, Intelligent
    Manufacturing and Robotics
King Fahd University of Petroleum
    and Minerals
Kingdom of Saudi Arabia

Kashif Nisar
University of Malaysia
Sabah, Malaysia

Muhammad S. Qaiser
Faculty of Mechanical Engineering
GIK Institute of Engineering Sciences
    and Technology
Topi, KP, Pakistan

Junaid Qayyum
Institute of Materials and Processes
School of Engineering, the University
    of Edinburgh
EH9 3JW, United Kingdom;
    Department of Mechanical
    Engineering
Institute of Space Technology (IST)
Islamabad, 44000, Pakistan

Syed Essa Rasan
Faculty of Mechanical Engineering
GIK Institute of Engineering Sciences
    and Technology
Topi, KP, Pakistan

Khalid Rehman
Faculty of Mechanical Engineering
GIK Institute of Engineering Sciences
    and Technology
Topi, KP, Pakistan

Roshan Rehman
Faculty of Mechanical Engineering
GIK Institute of Engineering Sciences
    and Technology
Topi, KP, Pakistan

Muhammad S. Sahar
Department of Mechanical Engineering
Khwaja Fareed University of Engineering
    & Information Technology
Rahim Yar Khan, Pakistan

Muhammad Umair Shafiq
Faculty of Mechanical Engineering
GIK Institute of Engineering Sciences
    and Technology
Topi, KP, Pakistan

Ali Akbar Shah
Mehran University of Engineering and
    Technology
Jamshoro, Pakistan

Muhammad F. Shah
Department of Mechanical
    Engineering
Khwaja Fareed University of Engineering
    & Information Technology
Rahim Yar Khan, Pakistan

Syed Waqar Shah
Department of Electrical Engineering
University of Engineering & Technology
Peshawar, Pakistan

Muhammad Zakir Shaikh
Mehran University of Engineering and
    Technology
Jamshoro, Pakistan

Adeel Tariq
Department of Mechanical Engineering
Universiti Teknologi PETRONAS
Perak, Malaysia

Shuhao Xu
College of Mechanical and Electrical
    Engineering
Nanjing University of Aeronautics and
    Astronautics
Nanjing, China

Omar Youssef
Mechanical Engineering Program
Middle East Technical University—
    Northern Cyprus Campus
Kalkanli, Guzelyurt, Mersin, Turkey

1 Promoting Enterprises of Functional Reverse Engineering through Design Thinking
Engineers' and Marketers' Perspective

Cedric Aimal Edwin

CONTENTS

DOI: 10.1201/9781003220985-1

1.1   BACKGROUND

The current global socio-economic environment is more dynamic and fluid than ever before. Local and international markets are anticipated to be more disruptive and unpredictable. The conventional methods and methodologies used to manage and promote enterprises have become redundant and proved insufficient. Developing countries have utilized principles of reverse engineering for centuries to deal with the unpredictable nature of the market. Functional reverse engineering (FRE), which mimics the functional features of an object to create either an accurate or enhanced virtual or physical model, allows enterprise owners to pace up innovation. Nevertheless, there is a dearth of available methodologies for FRE enterprises to promote their new ventures. This chapter argues that design thinking is a methodology that can be used to develop products or services of FRE and can also be deployed to promote enterprises of FRE.

1.2   FUNCTIONAL REVERSE ENGINEERING
AND DESIGN THINKING

Functional reverse engineering is a form of reverse engineering that focuses on the functions of an apparatus or system rather than its physical appearance. Enterprises of FRE perform activities of FRE to develop products, services, strategies, policies, procedures, processes, or materials. It has been shown to be beneficial in designing new systems. Design thinking is a design process that considers aspects such as empathy for the end-user and understanding how people interact with products, services, environments, and other people around them. This chapter will explore how innovators can promote enterprises of FRE through design thinking methods to help improve their design skills and create better products for their customers. Design thinking is being used more often by businesses because it helps identify problems before they happen by listening to customer feedback and then implementing solutions based on these insights while also taking care of essential human needs. FRE is a process that involves obtaining knowledge about how something works by taking it apart. Yet, FRE does not just involve breaking things down into their components but also figuring out what the original designers wanted to do with them and then using that information to inform original design decisions. FRE explains how a product or service is put together and all of the intricacies involved in its development. It also provides insight into why different parts are made from different materials or have specific shapes and what special considerations should be taken when designing new products based on this old one. Since FRE is often used to develop products which are not available in the market, thus, a unique approach to promote those products is required. It is argued that design thinking methodology can be adopted to promote enterprises of FRE.

Design thinking is the process of making decisions and strategies that are fit for today's evolving market. Design thinking has evolved as a methodology adaptable to changing market conditions, meaning it can be used by any business or organization looking for an edge in this rapidly changing world. Design thinking has evolved as a methodology adaptable to the changing market conditions and is commonly employed by businesses to design strategies, including but not limited to promotional strategies. Design thinking inspired organizations such as General Electric (GE), Google, and Apple to create systematic innovation for their customers, emphasizing people's needs rather than just looking at what would make them rich or how they could save money through these innovations. Design thinking aims to be more mindful about designing around customer pain points and focuses on reaching out while being empathetic towards those who need it most. This chapter argues that design thinking can be successfully employed to promote enterprises of FRE. What is design thinking? Why has it become so popular? How can it be used to promote enterprises of FRE? What is the new role of advertisers and marketers in this context? The following discussion will answer these questions and present a framework to promote enterprises of FRE in these turbulent times.

1.3 POPULARITY OF DESIGN THINKING

Design thinking helps to solve big, wicked problems in a short time. It is one of the many ways in which engineers and innovators can overcome complex problems. Design thinking is an iterative methodology that aims to understand users, redefine problems, and challenge hypotheses to explore alternate solutions and strategies that do not automatically appear in the initial understanding level. It is an approach that involves a collection of different hands-on approaches, for example, lightening decision jams, customer journey mapping, crazy eights, solution sketches (heat maps), storyboarding, and five-act interviews.

Design thinking is the methodology that helps to understand people for whom the goods and services are designed. It cultivates a strong interest in knowing them, so these products or services may more effectively address their needs. This knowledge of the end-user often allows the promotional team to develop strategies which are relevant and attractive to the end-user. Design thinking allows enterprise owners, engineers, managers, advertisers, and other stakeholders to challenge questions, hypotheses, and consequences. By re-enacting these issues and coupling a human-centered approach, innovators build multiple solutions in brainstorming sessions and take a practical approach in prototyping and testing phases, using this approach. Design thinking requires constant iteration and experimentation with innovative models and ideas, which is essential for promoting enterprises of FRE. By considering the right tools to start prototyping, innovators and marketers can get a sneak peek into the future and get a glimpse of how a specific promotional tool will perform with the users. Thus, instead of putting thousands of dollars in a particular promotional strategy and then analyzing its impact, design thinkers prototype promotional strategies on a small-scale and get a sneak peek into how it will be received by the end-user.

1.4 THE FIVE PHASES OF DESIGN THINKING

In the twenty-first century, there are different variations of design thinking. There are three to seven stages and modes in use, and different innovators deploy various versions of design thinking according to their requirements and product/service/

strategy development stage. However, the model presented by the Nobel Prize winner Herbert Simon in *The Sciences of the Artificial* in 1969 embodies all variants of design thinking. The Five Phases Paradigm from the Hasso-Plattner Design Institute at Stanford University (also known as the Stanford d.school) is at the center of the proposed framework [1].

The five stages of design thinking according to the d.school are:

1. Empathize (with your users)
2. Define (user needs, their problems, and your insights)
3. Ideate (by challenging assumptions and generating ideas for innovative solutions)
4. Prototype (begin creating solutions)
5. Test (solution)

It should be noted that these five steps will not all be linear and may be repeated iteratively in tandem.

The figure shows the cyclic process design thinking which begins with *empathizing* with the problems of the end-user; then the problems are *defined* properly so all team members have a clear and uncontested understanding of the problem; then the team generates innovative *ideas* (through the flare and concentrate process—discussed later in this chapter) to solve the problem at hand; after generating various ideas one or more ideas are *prototyped*; and finally the prototype is *tested* by the end-users and team observes how the end-user perceived or used the prototype. This process is repeated in cycles to solve complex problems in a short time.

Design thinking and FRE methodologies are used for designing new products or services to increase their marketability based on customer feedback and data analysis. Design thinking and FRE concepts are used daily by engineers for product development and product promotion. The following discussion highlights the different phases of design thinking and their relationship with FRE.

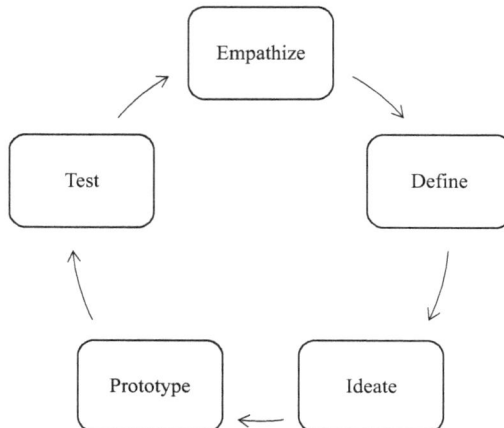

FIGURE 1.1   The figure shows the design thinking process.

1.4.1 EMPATHY (WITH USERS), DESIGN THINKING, AND FRE

Empathy is a word that has been used in the design field for many years. It is an essential tenet of human-centered design, and it can be applied to both personas and products. Design thinking and FRE utilize empathy to create successful designs, where products are taken apart to see how they work. When these two methods are combined, they are called "empathetic reverse engineering"—the process of taking a product apart piece by piece, understanding its functionality with an end goal of building something from scratch with superior value. Empathy is the ability to understand and share the feelings of another. In engineering, empathy can be acquired through design thinking and FRE to create more user-friendly products. For example, when designing a new product for an older person with arthritis, engineers should consider the difficulties they face daily and how their arthritis may affect them during the use of any device. This means considering factors such as weight, size, grip-ability, and flexibility, among others. Then, by understanding what they need from the product—such as something lightweight or small enough to fit in a bag—engineers will be able to build it accordingly so that it suits their needs better than existing solutions do. Whereas marketers use the same information to develop promotional strategies by highlighting the features that propped in the empathy exercises. Design thinking is a user-centered method of designing, creating, and making that employs empathy with users to understand their needs.

Usually facilitated by a design team, product development and promotional strategies typically include brainstorming about potential solutions to meet those needs, prototype testing, and refinement until it satisfies the needs. The engineering design process and promotional strategy development process are cyclical in nature. Engineers and marketers iterate through phases of creativity, experimentation, and evaluation to produce an artifact that meets the needs of its users. Empathy can be used in conjunction with design thinking and reverse engineering as critical components to creating innovative promotional solutions grounded in understanding user needs. In the empathy stage of design thinking, marketers interview end-users and form shadowing groups to observe the behavior of the end-user. Throughout this stage, the aim of the marketers is to seek understanding and be non-judgmental. Marketers are trained to be analytical, efficient, and logical in their approach. They use problem-solving approaches such as reverse engineering to problem solve when they face technical difficulties. Marketers often employ the design thinking process to create new promotional strategies that are more appealing to the consumers. Empathy is a powerful tool used by marketers (and promotional experts) in the design thinking process, allowing them to understand better the needs of those who will use their product or service, and then design marketing content which highlights the needs and wants of the end-users. With empathy comes an understanding of how people feel about certain things, what motivates them and why they behave in specific ways. Empathy can help marketers develop promotional strategies that people want and need while also creating a strong sense of connection with potential customers through designing experiences that make users feel good about using the specific product.

1.4.2 DEFINE (USER NEEDS AND PROBLEMS), DESIGN THINKING, AND FRE

Engineers and marketers are often faced with a problem that they need to solve efficiently and effectively. An engineer might solve this problem by defining the user needs and problems, and through design thinking, figuring out how to build the solution. Whereas a marketer solves the promotional strategy conundrum by defining the user needs and problems. FRE is the technique engineers and marketers use when they want to see how something was designed or made. The goal of FRE is not necessarily solving an existing problem but rather understanding what went into designing or making something to learn from it for future projects. Engineers and marketers have the skills to take complicated problems and turn them into elegant solutions. That is why they are always needed in any organization that deals with complex products and processes. The design process can be efficient even without an expert on hand by implementing specific design thinking rules. The key is to identify user needs and reverse engineer their experience using design thinking principles. By properly and clearly defining the user needs and problems, engineers and marketers can create better designs faster than ever before!

Engineers and marketers are often required to design a solution for an idea before the need is fully understood. This can be difficult when there is limited knowledge of the problem, and no understanding of what customers want. Design thinking has been used in various industries to help bridge this gap by focusing on empathy, exploration, and innovation. FRE is a process that helps identify user needs through observation, experimentation, and collaboration with users. By combining these two practices, a better understanding of customer's problems and their desired solutions is achieved while developing a strategy for addressing them effectively. In the *define* stage, marketers develop user persons, clearly outline role objectives, discuss challenges and decisions, and highlight pain points. When these data form a part of the promotional strategy, the promotional message is such which appeals to the emotions of the end-users.

1.4.3 IDEATE (GENERATE INNOVATIVE IDEAS), DESIGN THINKING, AND FRE

Engineers and marketers are in the business of solving problems, and this is what they do best. When faced with a problem, engineers and marketers will use one or more of their skills to solve it. Sometimes they have to think about how something works before they can make it work better. They often start by brainstorming ideas and then choose which ones are worth exploring further. For example, when designing an airplane wing, engineers had to figure out how air moves around the wings (ideate) so that they could design them for greater efficiency (design thinking). Similarly, while designing a promotional strategy for an airline, marketers had to figure out how end-users will react to different kind of marketing content (ideate) so that they could design them for greater efficiency (design thinking). With FRE, engineers and marketers take something apart (a tangible product or an intangible marketing strategy) in order to understand how it was made and why certain decisions were made during formulation (reverse engineering). Engineers and marketers use these same skills every time they need to solve a complex problem. Similarly, marketers also use design thinking and FRE to develop unique strategies for promotion.

On one hand, design thinking allows marketers to develop promotional strategies which are grounded in real data and tap into the real needs of the end-users, and on the other hand, FRE allows marketers to take apart the promotional strategy utilized by competitors, analyze it, and customize it according to their own needs. In the *ideate* stage, marketers use various techniques of sharing ideas, deploy the diverge/converge approach, and prioritize ideas based on relevance, importance, and impact.

1.4.4 PROTOTYPE (CREATING SOLUTIONS), DESIGN THINKING, AND FRE

Engineers and marketers are faced with many different challenges in their work. These can range from creating a new prototype to designing solutions for an existing product or process or strategy or reverse engineering an older machine. There is never one good solution that fits every engineer's or marketer's needs, but a few common approaches have been proven and tested over time. Engineers and marketers are often required to prototype new solutions to demonstrate through empirical evidence that their solution works in the real world. Engineers and marketers work with prototypes regularly to create solutions that may never have been thought of before until they were prototyped.

Designing a new product is not just about coming up with an idea and drawing it out. There are many other ways to go from initial concept to final design, one of which is prototyping. Prototyping can be done in many ways, such as using hand-written drafts or clay models or 3D printing. Though, it can also be used to test idea before starting to make anything substantial. This way time, money, materials, and other resources are not wasted. The first step of any good prototype is understanding how the product will function and what its intended purpose is—this process is the core activity within FRE. Once this step has been completed, prototypes can then be made accordingly, so they mimic the original.

Prototyping is an essential tool for engineers and marketers. It provides a way to test and validate design and function decisions before making costly commitments in terms of time, money, or effort. Prototypes can also be used as working models to demonstrate the benefits of a design or process or strategy to potential customers. There are many different methods for designing prototypes, such as sketching out ideas on paper, 3D printing, rapid prototyping with laser cutters and CNC machines, model-making from clay or other materials, and more. The most important thing is not which method is used but the idea behind it: make sure the prototype will allow the developers to test any assumptions about how the invention works at every stage of development from initial concept through final product rollout.

1.4.5 TEST, (SOLUTIONS), DESIGN THINKING, AND FRE

Engineers, marketers, designers, and other creatives must analyze a product's function to design a solution. This is often done through FRE—taking apart the product to understand how it works. Engineers are faced with many challenges when designing an object while marketers are faced with challenges while promoting that object. They must focus on the design and testing to ensure that they meet safety standards and be mindful of how the design could break down in use. Design thinking is a process that helps engineers and marketers develop innovative solutions by understanding their users' needs and constraints, while FRE allows engineers and marketers to explore existing designs for new opportunities. Engineers design and build things. Marketers promote those things by designing appealing content for the end-user. Engineers and marketers often encounter designs that are not well thought out or have errors in them. Sometimes a quick fix is all it takes to get what is needed for the completion of the project. Other times, the best solution is to functionally reverse engineer the object or strategy to figure out how it works so that it can be improved upon it with new features or better engineering/marketing principles.

Engineers and marketers are a valuable and integral part of the product development team. They provide solutions to problems, design new products, and perform complex tests. They use design thinking, a systematic approach to problem-solving that includes empathy for the people who will be using their solutions and an understanding of what is needed for the solution to work. FRE is an additional technique engineers and marketers can employ when designing products, services, processes, or strategies. It involves breaking down a product, service, process, or strategy into its component parts so they can be analyzed and understood, then reassembling them in a way that solves the problem at hand.

1.5   THE NEW ROLE OF MARKETERS

Chief marketing officers (CMOs) play a role in promoting FRE enterprises changes from campaigners to customers. Marketers do not need only fuel market expansion but change business entirely. To generate revenue, CMOs must lead their teams and expand beyond traditional marketing. CMOs ought to step up their lead and accelerate creativity through their consumer experience. One of the biggest challenges for CMOs is creating effective marketing campaigns. Nonetheless, there are a few things they can do to improve creativity and drive results. Marketers are increasingly becoming more focused on gaining more insight into the customer journey, and the tools and technology that help us do this are essential.

CMOs are considering moving away from the status quo and facilitating disruptive innovation. The CMOs and other marketing professionals are focusing more on how to facilitate disruptive innovation. The CMOs are concentrating their efforts on how to enable disruption in the marketplace. This is a significant change because they will be able to bring new ideas without losing its core values. Consequently, CMOs will be able to bring new ideas without losing the company's core values if they know how to manage that process. Under the leadership of an exemplary CMO, companies will be able to develop and launch new products and services in their existing markets.

1.6   MARKETERS LEADING DISRUPTIVE INNOVATION

Innovators are the individuals who create new products, technologies, or business models to disrupt an entire industry. While marketers ensure that the information regarding the new product, service, technology, or business model ignites interest, curiosity, and action among the end-users. Disruptive innovation makes incumbent innovation obsolete and changes the consumption pattern of the customers. In the initial stages, disruptive technologies usually contain glitches and bugs, but they have the potential to be market leaders. The challenge for marketers leading disruptive innovation is not only in creating quality marketing and promotional campaigns but also in identifying the right market segment. Marketers often meet this challenge with creativity and innovation to lead their organizations into success through disruptive innovation. Creativity and innovation in marketing and promotional campaigns are a result of the design thinking approach.

Innovation has become synonymous with success, even though it can be challenging to identify when an organization's next breakthrough idea might come from within. Disruptive innovation makes the incumbent technology, strategy, or policy irrelevant and "disrupts" the consumption/usage pattern of the end-users. For disruptive innovation to pass through its rudimentary stages, marketers need to play a leading role. It can be argued that due to the efforts of marketers, disruptive innovation is able to gain enough traction that it is considered seriously by the market. Disruptive creativity and innovation call for leadership bravery and conviction. It involves blurring the gaps between commercialization, distribution, products, and new technologies and concentrating on actual consumer value, and getting the necessary resources to render the possibility of converting disruptive innovation into a new norm.

1.7   DESIGN THINKING AND PROMOTERS OF ENTERPRISES

Design thinking is a process that uses empathy and creativity to help businesses innovate. The focus of design thinking is not on what is made but how is it made. This perspective allows for new perspectives in the way that business is conducted. Design thinking is a process whereby an individual or group thinks deeply about all aspects of a situation from different angles to find solutions for problems that no one has yet solved. Designers can be individuals or groups who systematically think about people's needs and desires, then brainstorm ideas and choose which ones work the best. From this viewpoint, engineers, and marketers, both, are designers. It has been found that design thinking is more effective than other methods of problem-solving when tackling complex, ill-defined or high-risk problems. Design thinkers also can communicate their ideas in ways that are easily understood by others so as to promote change and innovation.

Design thinkers can be found in every industry, but there is an increasing demand for designers within organizations who specialize in product development and marketing communications jobs. These individuals need to be able to think creatively about new solutions and know how to articulate those thoughts with clarity and detail to influence others' understanding of the work they do on behalf of their organization.

Effective communication strategies bind and touch hearts emotionally. Promoters and marketers of enterprises are becoming less concerned with the financial aspects of their strategies and focusing more on connecting and building relationships with the end-users. Engineers and promoters of enterprises of FRE connect with future clients with qualitative data from design analysis, aside from the objective market research figures. This helps innovators consider those end-users purchasing products of FRE based on its contribution to their facility, rather than how much the consumers will pay for the products/services.

Another critical part of the design thinking methodology is developing a narrative. Promoters of enterprises of FRE focus on functionality, design the messaging around human needs, and engage the customer on a more personal basis. Promotional content that makes consumers chuckle or weep is far more influential than those mainly about product characteristics and cost savings.

1.8 DEVELOPING PROMOTIONAL STRATEGIES FOR SMES USING DESIGN THINKING

In order to compete in today's marketplace, small and medium-sized enterprises (SMEs) need to use innovative promotional strategies. Design thinking provides an invaluable toolkit that SMEs can utilize to create successful marketing campaigns. Designers can develop a more sustainable, cost-effective approach when working with SMEs and provide resources for those looking to start their own business. Designers are problem solvers who work at the intersection of creativity and innovation. They can bring about change by designing solutions that meet people's needs while also being environmentally conscious and socially responsible. They are uniquely qualified to help SMEs find creative ways of promoting their products or services.

Designing is a process of problem-solving that requires creativity and critical thinking. Design thinking is an idea-generating technique that can be applied to solve any business challenge, from marketing to communications. This method encourages creativity and innovation by bringing together people with diverse skill-sets, facilitating collaboration through design workshops, designing prototypes to test new ideas or methods, and finally evaluating if the solution solves the initial challenge. SMEs can use design thinking as their promotional strategy to foster innovation within their company's culture. The use of design thinking is becoming more prevalent in the business world due to the changing socio-economic and environmental circumstances. SMEs can use design thinking as an effective strategy for generating innovative promotional strategies. Design thinking, through its various lenses, generates different types of ideas which can then be used as bases for developing promotional strategies. Companies using design thinking techniques create successful marketing campaigns with very different audiences (from targeting senior citizens to the other college students; from B2B customers to B2C).

The following are some of the salient features of promotional strategies of enterprises of FRE.

1.8.1 TAKE THE END-USER FIRST APPROACH TO PROBLEM-SOLVING

Modern problem solving is often seen as a linear process of analysis, research, and decision-making. This approach is inefficient and ineffective in the face of ambiguity that arises when confronted with new challenges. To be successful at solving problems, an end-user first approach is required which reverses the linear process by starting with research on the user's needs and then analyzing and designing a solution for them before finally making our decision. The end-user first approach to problem-solving states the best way to solve a problem is by understanding what the user needs and wants. The end-user first approach was developed to shift focus from making products better for producers but making them better for users of those products. This approach helps innovators design new solutions that make life easier for their customers instead of adapting their lives around the product. Major disruptors of the last two decades, Uber, Amazon, Alibaba, Airbnb, etc., have the same fundamental characteristics of success. All these disruptors acknowledged and addressed their client pain-points and frustrations with their marketing and promotional content. Marketers often rely on product characteristics to attract and maintain consumers and fail to ignore the problems they wish to fix. Empathy is the secret to solving genuine issues for consumers. Innovators, engineers, and marketers must be able to listen, observe and immerse themselves in user experience. This implies that enterprise founders and promoters need to come out of their ivory towers, get rid of the quantitative survey, and directly engage with clients and frontline teams. Innovators, engineers, marketers, and promoters need to think of clients beyond financial figures, and they must get involved in the dreams, worries, emotions, and unique backgrounds of their clients. Enterprises of FRE can conceive products, services, strategies, and policies in an entirely different manner once they effectively empathize with their clients. Various design thinking tools can be utilized to tap into the pain-points of end-users, such as, empathy mapping, journey maps, drafting buyer personas etc. Promoters of enterprises of FRE need to prioritize resources and budgets to methodically maneuver user experience.

1.8.2 Free the Artistic Genius

In the business world, it is often thought that an artist needs to be willing to compromise their work to produce something that will sell. This is not true, especially for promoters of enterprises of FRE. The role of the artist has been seen as a luxury. It is time for this idea to change and to free the artist! Creativity is often associated with the ability to create something new. Freeing an artist's creative genius requires a lot more than simply giving them freedom. Studies show that creativity levels are hindered by stress and sleep deprivation, both of which can result from an unsupportive environment or lifestyle choices. For marketers to flourish in their careers, they need consistent support: financial, emotional, social, and intellectual, all of which allow them to produce their best work without fear of judgment or criticism. Imagine what could be achieved if it were easier for artists to be productive! Creativity and innovation are crucial to the success of enterprises of FRE. To be innovative, marketers need a space where they feel free to create without fear of judgment or criticism. Marketers within an enterprise must be encouraged with resources and opportunities to foster their creativity through education, mentorship, and collaboration. Companies need to invest in marketers' well-being by providing them with time off so that they can explore their creativity outside of work. Creativity in tandem with insight is the foundation of creative invention. The in-house team develops this opportunity to put versatility into the company to reimagine the potential of new data and technologies. Indeed, most disrupters did not develop the technologies they succeeded but found an easy, new and inventive way of solving a human dilemma.

1.8.3 FLARE AND CONCENTRATE

With each discussion phase of design thinking there are further two stages: flare (the brainstorming stage) and concentrate (the refinement stage), are critical features for achieving innovation in the promotional strategy of an enterprise of FRE. By considering these two essential stages and how they affect the outcome from an entrepreneur's perspective, the creative effectiveness of the promotional strategy of enterprise of FRE is paced up. Designers need to seek inspiration from different sources and be very aware of what is happening around them at any given time. It could lead to a new idea or solution for their problem. These two elements, flare, and concentrate, help marketers find a balance between both ends of the spectrum so they can create something unique and original! Design is used to test theories and to find possibilities through flaring approaches or differing reasoning. It consists of collecting a broad range of consumer stories to understand opportunities and quantity better and delay judgment. Using attention or convergent reasoning strategies to limit the options for educated choices until there are a variety of ideas or issues is a successful strategy to deploy. When the imaginative juices spill through to the completion of a dream, marketers know precisely when to focus and when to explore ideas from other industries. Design thinking is not just a creative process but also an analytical one. Marketers must first assess the needs of their audience and then decide on the best way to solve them. Flare is required to explore different possibilities before deciding which one will work best; concentrating helps narrow down what design decisions need to be made to make it happen. Marketers who utilize these strategies effectively can quickly save time and money by making more informed design decisions without sacrificing quality or creativity. These tactics are especially relevant when designing with limited resources as they require less investment upfront yet still have high potential return rates later in the design.

1.8.4 UNDERSTAND AND KEEP GOING FROM STUDIES

There is no acceleration in conventional marketing processes. It is impossible to step away from a strategy or decision after resources and dollars have been invested into a concept. A good promotional strategy for enterprise of FRE, balances data-driven approach and intuition. Often it is a good idea to go away from empirical data to generate out-of-the-box ideas. A positive attitude towards experiments without remorse to speed up the process is often found very useful. Because ideas are cheap, the purpose is to explore them, uncover the issues and develop them. The concept of *Design Sprints* is valuable and apt for FRE because it allows time-bound innovation by quickly pushing the innovators through the test, learning, and iteration stages. Innovators go to market for testing and learning for soft releases, pilots, and trial models. The innovators also save resources by investigating, briefing, and exploring

TABLE 1.1

Recommended Promotional Strategies for Enterprises of FRE

Product Development Stage	Suggested Promotional Strategies
Pre-release	Newsletters; presentations; videos; trade expos
Release	Demos; press releases; tutorials; events
Post-release	Testimonials; case studies; referral marketing

Adapted from: Dunford, A., 2019 [2]. *Obviously Awesome: How to nail product positioning so customers get it, buy it, love it.* UK: Derbyshire, United Kingdom, DE11 8LN

suggestions that do not suit the clients. This helps get a better understanding of prosperity, size, and new possibilities.

1.8.4.1 Pre-Release Promotional Strategies

In the pre-release stage, enterprises of FRE can use *newsletters* to promote their products. This will help them get feedback on how consumers like it and what features they would want. It also helps with getting feedback for any bugs that are found in the product before its release date. Emails are usually sent out to a subset of people who have signed up for this service or were invited by someone else who has signed up, so the marketers know that there is an interest in early access into these emails from the enterprise. This is done through carefully curated content and a personalized experience for each subscriber. The company newsletter provides subscribers with up-to-date information on industry trends, as well as news about the companies themselves. They are also used to make announcements such as new product launches or discounts and promotions. A corporate blog post will often have links in it that let readers know where they can find more information about the topic discussed in detail within the article itself. For example, Dell used newsletters as part of an experimental marketing campaign for its new laptop models in 2015. The company was able to test different kinds of content such as design principles; target audiences; and delivery methods before launching the product on a large scale. This way, they were able to find what works best for them before committing too much time or resources into it.

In the pre-release stage, enterprises of FRE can also utilize *presentation* to promote their product. Presentation is a key component of marketing, and it is important for enterprises not only to create an image for themselves but also to maintain that image by ensuring that they are presenting the best content possible. Presentation is a type of communication that relies on the spoken, and written word to convey information. It has been used since ancient Greece as an art form in public speaking. Pre-release presentations are often created to present new product/service concepts or ideas, but they can also be used for other purposes such as demonstrations and lectures where the knowledge from one person is shared with an audience. Presentations are a tool that enterprises use in the pre-release stage because it helps them promote their product by creating interesting content for potential investors who might not know about them yet. For example, GE presents how they are using data analytics and visualization to make better business decisions. These examples show that enterprise innovation is more than just a buzzword; it's an ongoing process of discovery.

In the pre-release stage, enterprises of FRE are also recommended to use *videos* to promote their products. A video is a powerful tool in marketing and advertising because it offers something that no other media does: immersion. For instance, if an enterprise that has just launched a new product but does not have any reviews yet, the video might be the best bet for getting the word out about what people experience when they use it. Enterprises of FRE can use emotions or social cues to advertise their products with videos to create an immersive experience. This is an effective form of marketing because it allows users to see what the product does and how they might be able to make use of it in their own lives. One such company that has done this well is Tesla. Enterprises are now using videos to promote their products in the pre-release stage. This strategy has been shown to be more effective than traditional marketing techniques such as print ads and billboards.

In the pre-release stage, enterprises of FRE are also encouraged to attend *trade expos* promote their products. Companies in this phase are usually looking for feedback on their products and want to see if they have a viable product to offer. A trade expo is a convention that allows innovators in a specific industry to showcase their products and services. There are many different types of trade expos, including but not limited to: technology, automobile, food service, and apparel. All these industries have unique needs that must be met when creating marketing campaigns for this type of event. Trade expos are becoming more popular in today's business world because they provide opportunities for companies to reach new markets with innovative product or service offerings while also providing an opportunity for customers to see what is available on the market without leaving the home ground. A key benefit of trade expos is that they offer exhibitors an opportunity to showcase themselves as they prepare for launch.

1.8.4.2   Release Promotional Strategies

In a world where innovation is the key, enterprises are looking for ways to promote their product. One of these methods is *demos*. Demos give people an opportunity to get a feel for what the product does and if it can meet their needs. Enterprises of FRE can use this method in the release stage as well as after launch when they need feedback on how to improve the product. This is a process where companies showcase their new product or service to generate interest and demand for it. There are also physical demos where people go into stores or speak with company representatives who show them how the product works in person. Enterprises of FRE engineering should create a demo that is as close to the real world as possible and consider what will be seen by the audience. For instance, if they are promoting a facial recognition system for law enforcement agencies, then it would be important to show how accurate it is in different lighting conditions and at various distances from the camera. It is also important not to make any assumptions about what knowledge attendees have about the product or service when designing an interactive demo because this could cause confusion for them while trying out the product or service during the demonstration. To get feedback on whether they should include these features in their future designs, enterprises are encouraged to conduct qualitative research with customers.

In the release stage, enterprises of FRE may use *press releases* to promote their products. Press releases are a way of communicating with journalists and media outlets about new product launches. They are released by companies to generate interest in the company's latest innovations or developments and can be used as an effective marketing tool for generating sales. These methods may include generating customer testimonials or featuring guest bloggers who have been personally impacted by their product. It is equally important that enterprises maintain good relationships with competitors and non-customers.

At the release stage, enterprises of FRE are recommended to develop *tutorials* to promote their products. For example, a company may provide instructions on how to assemble a product that is being released for sale. This will help customers understand how to best operate the new product and may also be used as an advertisement for the company's other products. Tutorials provide a way to get word out about new features and functions of an app or product. Enterprise companies are getting smart about this strategy—Apple has created a video tutorial series called "Introduction to Apple Pay", which is available on YouTube as well as the Apple website, and they have other tutorial videos for each of their apps. The company plans to add more tutorials in response to customer feedback over time. Tutorials are a great way for companies to communicate complex ideas and offer customer support. They are also an excellent resource for customers who want to learn more about a company's product or service but do not have time to call or email them with questions.

In the release stage, enterprises of FRE are encouraged to attended relevant *events* to promote their products and generate buzz. There are many different types of events that enterprises can choose from, but trade shows and conferences are the most common for enterprises of FRE. *Trade shows* are a great way for enterprises to showcase new or existing products in front of large amounts of people. They allow companies to interact with potential consumers and customers face-to-face. These events also allow companies to see what other competitors have been doing. *Conferences* offer an even more targeted method for enterprises that need publicity but would prefer not to be in the spotlight as much as they would at a trade show; this is because it's usually smaller scale than a trade show.

1.8.4.3   Post-Release Promotional Strategies

Enterprises continuously strive to innovate and provide their customers with the best product. In the post-release stage, Enterprises of FRE are recommended to develop *testimonials* of end-users to promote their product. These testimonials are a great way of showing consumers that they have already successfully used this product to get them on board. There is an important aspect of these testimonials: being authentic. Consumers want real reviews from people who have tried out the products, not just someone telling them what they think it is like. Video testimonials have a higher authenticity perception. It would be better if there were some sort of rating system that could help determine which reviews are genuine or not to weed out fake ones and give consumers more confidence in buying from reputable companies. Enterprises must be authentic and show real value propositions as well as provide rich customer reviews or testimonials from those who are using the products/services.

The post-release stage is the most crucial time for enterprise generally, but more so for enterprises of FRE. This is enterprises of FRE they can promote their product with *case studies* to continue building their reputation and customer base. Post-release case studies give insight into how a product has been used in real-world settings which gives companies an opportunity to tweak their marketing strategy or make design changes that will improve sales and customer satisfaction. The release of a new product is just the start. Once the initial excitement has died down, enterprises of FRE need to focus on long-term success and sustainability. One way to do this is by using case studies that promote their products, which can be used for marketing and recruitment purposes. Case studies are also helpful in identifying areas for improvement or what was successful post-release. The data compiled from these case studies should be analyzed thoroughly to make sure it's accurate and up to date so that it can provide insights into future releases.

Post-release is a critical stage in product life cycle management because it is when the business has to gain brand recognition and market share. One way for companies to achieve this is by using *referral marketing*, which requires them to rely on word-of-mouth from customers who are satisfied with the company's products or services. This strategy relies on a psychological principle known as "social proof" where people assume that if others have done something, then there must be some value in it. The post-release stage can be very challenging for enterprises of FRE because they may not yet have sufficient brand recognition and consumers might not know about their products or services. Referral marketing provides an opportunity for these enterprises of FRE to use social proof while also establishing trust between themselves and potential customers. By using this method, companies can get customers and potential customers to refer others by providing incentives for doing so. These incentives can be anything from discounts on a purchase, free items with a purchase, or even cash rewards for successful referrals. The idea behind referral marketing is that it encourages people who have already used the product or service before to tell someone else about what they do and share how happy they were with the experience.

1.9 CONCLUSION

To promote enterprises of FRE, design thinking methodology was found to have characteristics and features which enhance systematic creativity and foster an end-user-first approach. Products, services, strategies, policies, procedures, processes, and materials developed through FRE are not only unique and novel but also lack public awareness. Design thinking is an umbrella term that encompasses various techniques, processes, and tools used to generate creative solutions. Design thinkers approach a problem by empathizing with the users' needs and then generating many possible solutions based on creativity and intuition rather than logical analysis and judgment. This method has been applied successfully in various industries, including advertising.

Design thinking makes it possible for marketers, promoters, or advertisers to live up to highly knowledgeable consumers' needs and develop a 360-degree marketing campaign. The Design thinking methodology addresses problem-solving progressively and tests results relentlessly from dollars to consumer loyalty standards. It is an innovation-driven approach that inspires new thought and develops insightful concepts but remains practical. It blends the appetite of consumers with what is theoretically practical for companies and commercially feasible. Design thinking is a process that makes it possible for advertisers to launch complex products and services. It consists of four phases: empathize, define, ideate, and prototype. The Design thinking process starts with empathy. The idea behind this phase is to understand what people are feeling or experiencing in their environment before making any changes or launching new initiatives. After this step is complete, the next phase involves defining which problems need to be solved or questions answered by coming up with tangible solutions (defining). Next comes coming up with ideas through brainstorming techniques such as clustering (ideating) followed by prototyping potential solutions from the previous step. Design thinking is a human-centered problem-solving approach that moves away from traditional linear approaches to innovation and problem-solving. It is an iterative strategy for identifying and understanding customer needs and then creating innovative solutions with customers. Through deploying the Design thinking process, promotional strategies, and tools (for three phases of product life cycle: pre-release, release and post-release) were developed. These strategies and tools will enable marketers to promote enterprises of FRE in a novel and creative way.

REFERENCES

[1] Design Thinking. (2021). Retrieved 24–08–2021, from www.ideo.com/pages/design-thinking

[2] Dunford, A. (2019). *Obviously Awesome: How to nail product positioning so customers get it, buy it, love it.* UK: Derbyshire, United Kingdom, DE11 8LN

2 Everything You Need to Know about Intelligent Manufacturing
Industry 4.0

Ali Akbar Shah, B. S. Chowdhry,
Tanweer Hussain, Kashif Nisar,
Muhammad Zakir Shaikh and Saifullah Samo

CONTENTS

DOI: 10.1201/9781003220985-2

2.1 INTRODUCTION

Adapting cutting-edge technologies has become a prerequisite for industrial and technological survival. There is a significant contrast between modern industries and the First Industrial Revolution (as seen in Figure 2.1). From steam-powered machinery to programmable logic controllers (PLCs), and from PLCs to more current systems such as remote telemetry units, companies are eager to adopt new and improved technologies that can reduce or eliminate downtime [1]. It is as a result of this that we have discovered a way to automate manufacturing processes in order to bring simplicity to users so that they can execute their duties more efficiently [2]. To understand the evolution of the industrial revolution, its timeline is stated as follows:

2.1.1 INDUSTRY 1.0

The introduction of steam-powered machinery and production automation in the eighteenth century ushered in the First Industrial Revolution. The mechanized version, which previously created threads on basic spinning wheels, produced eight times the volume in the same amount of time. The power of steam was already well-known. The greatest breakthrough in enhancing human productivity was the use of information technology for industrial applications. Steam engines could be employed to power weaving looms instead of physical prowess. Further substantial changes occurred as a result of innovations like the steamship and (about a century later) the steam-powered locomotive, which allowed people and freight to travel long distances in less time.

2.1.2 INDUSTRY 2.0

The Second Industrial Revolution began in the nineteenth century with the development of electricity and factory line manufacturing. Henry Ford (1863–1947) got the concept for mass manufacturing from a slaughterhouse in Chicago, where pigs were

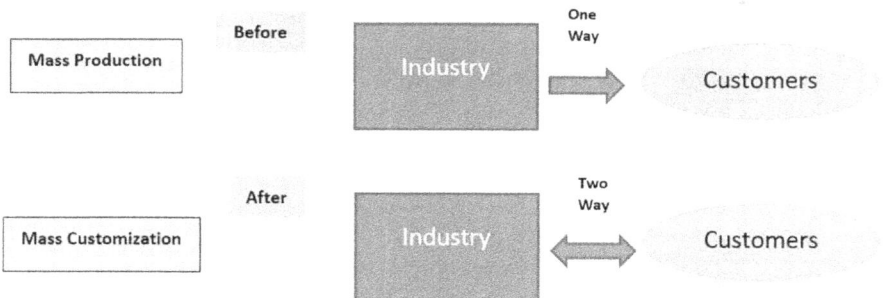

FIGURE 2.1   Mass Customization [5].

suspended from conveyor belts and each butcher was only responsible for a portion of the slaughtering process. Henry Ford applied these concepts to the automobile industry, transforming it significantly in the process. Previously, a full automobile was built at a single station; now, automobiles are built in partial phases on a conveyor belt, which is far faster and cheaper.

2.1.3 INDUSTRY 3.0

The Third Industrial Revolution, often known as the Digital Revolution, began in the twentieth century and was characterised by extensive use of electronics and computers, as well as the introduction of the Internet and the discovery of nuclear energy. This age saw the growth of electronics like never before, from computers to new technologies that enable the automation of industrial activities. Advances in telecommunications set the way for widespread globalisation, allowing businesses to relocate manufacturing to low-cost economies and radicalise business models all over the world.

Industry 4.0, also known as the Fourth Industrial Revolution (IR 4.0), describes a new set of organisational levels with strict control throughout the whole product life cycle [3]. Because it values customer feedback, it shapes its final product to meet customer requirements so that the end user is satisfied, as seen in figure 2.1. Consequently, the industry operates efficiently, as it encompasses research, commissioning manufacturing processes, on-time product delivery, and product recycling (see figure 2.2). Industry 4.0 encourages machine-to-machine communication as well as internet access (on a primary server or in the cloud) [4].

Moreover, the product modeling systems are monitored and assessed because they are developed as per customer requirements [6]. The entire manufacturing processes are segmented into many steps and each individual process behaves as a node. These nodes are able to communicate with one another and to the main server. This aids the correlation of multiple production processes, resulting in reduced delay time [7].

FIGURE 2.2 Faster Production for Reducing the Delay Time [8].

The major goal of Industry 4.0 is to transform current machines into AI-based self-adaptive machines in order to improve predictive maintenance and identify any mechanical defect before it occurs. Sensory nodes must communicate with one another and with the main server in order to perform predictive maintenance. It is so because the real-time data monitoring could be carried out and industrial harm might be identified quickly.

2.2   NINE PILLARS OF INDUSTRY 4.0

In Industry 4.0, data collection is crucial as its model is completely based on predictive maintenance. It uses the data available in the community support from the entire industrial communities for the logistics so that data can be collected for the faults in order to avoid the downtime. Industry 4.0 modelled on 9 main pillars as shown in Figure 2.3 [2]. This helps in transforming the mass production industry into a customized one. For further customizing any existing industry as per Industry 4.0 infrastructure [9].

2.2.1   DATA ACQUISITION

The most important pillar out of the nine pillars is the data collections. The data can be collected from various sources like various communities, sensors, databases or from cloud services.

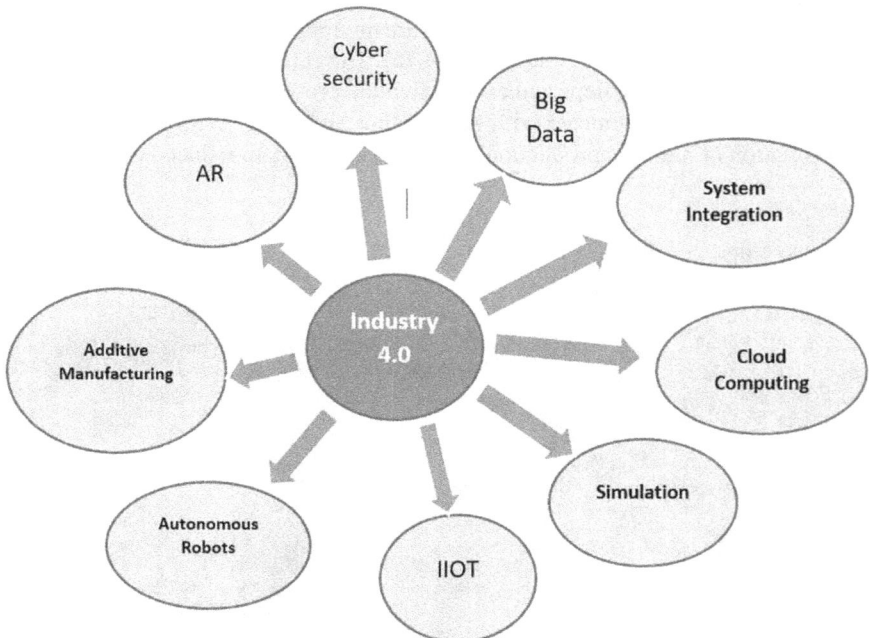

FIGURE 2.3   Nine Pillars of Industry 4.0 [10].

The data collected can be used for the predictive maintenance allowing machinery fault could be diagnosed and tracked in a real time environment. The major advantage of real time fault detection is that actions for mitigating the industrial fault can be taken in a timely manner to further reduce the downtime [11].

2.2.1.1 Product Life Cycle of Data Acquisition

If considering the product life cycle of the data acquisition, it comprises of five parts:

- Data capturing: Acquiring the data from a reliable source.
- Data storage: Storing the data in a secure memory source.
- Data usage: Data can be used conveniently (for e.g., can be read, write or etc.).
- Data archival: It is a way to store data in a secure location so it can be secured and easily recovered.
- Data destruction: To remove un-necessary data from every source that is saved into.

2.2.2 ROBOTS

Humans have a proclivity towards making mistakes. They can collaborate with autonomous robots to decrease these mistakes. Robots are advancing at the same time as technology [9]. Robots are now more precise and accurate than they have ever been. They are intelligent and calculated, and they can execute perilous activities that people cannot do it easily, safely, correctly and cooperatively with human employees [10–12].

2.2.2.1 Product Life Cycle of Robots

Product life cycle of robots is dependent on its working conditions in which it operates in industry. According to various statistics, it is known to us that an industrial robot lasts for 80,000 to 100,000 [13]. Which is typically six to eight years' service life.

2.2.3 SIMULATION

Simulation has given rise to technologies such as the digital twin model. Simulation is a technique for establishing a virtual workspace in computers that allows any issue to be identified using 3D modelling [11]. Furthermore, 3D models of bespoke items may be developed so that they can be inspected using various ways [[12,14]. For example, if we want to examine a product, we may virtually generate it on ANSYS and then evaluate its concerns linked to design parameters utilising techniques such as finite element analysis (FEA).

2.2.3.1 Life Cycle of Simulation

The life cycle of simulation is mainly based on the data collection. Then according to the data, the model can be built as per industrial needs. The life cycle of the simulation is robust as much as the data acquired is relevant to the model. Given that, the model is static and not dynamic because in dynamic nature of simulation, the precision of the instrumentation is compromised.

2.2.4 SYSTEM COMMUNICATION

Unlike traditional industries, Industry 4.0 is equipped with communications as mentioned as follow [2,9]:

- It involves sensory communication between each and every sensor.
- It has process to process communication.
- It has access to communicate with the main server.

Whereas, in the traditional industrial ecosystem, only few devices were allowed to communicate with the main server that caused complexity in the industrial fault diagnosis and tracking.

2.2.4.1 Life Cycle of System Communication

One of the most important aspects of civilization is communication. People need each other, and in order to engage, they must be able to communicate and comprehend each other. The communication cycle describes how the mechanism for transmitting and receiving messages works. As long as the industrial devices communicate as per international standards, the communication devices will stay relevant.

2.2.5 INDUSTRIAL IoT

Industrial IoT (IIoT) is a protocol based on the internet of things (IoT) [15] that allows connectivity and communication between makes industrial devices. Due to this, it has vast applications as elaborated in Figure 2.4. The connectivity of these devices is dynamic in nature [8]. The data transmitted is acquired from various manufacturing processes for enhancing the production environment. The sensory units mounted with each, and every manufacturing process are able to communicate with one another and with the main server. This helps the industry to identify any fault in a real time environment. Moreover, these sensors are smart and are standalone. So,

FIGURE 2.4 Applications of Industrial IoT [15].

therefore based on the data available in the form of libraries, these sensors can be trained using algorithms based on deep learning and big data which helps them to identify any fault on their own [23–24].

Moreover, due to emergence of unsupervised learning, these sensors can also identify any new fault which is not even recorded before. In order to validate the unknown identified faults, the sensors can communicate with one another and can track down the very start of the fault. In this way, the entire affected areas of the industry are easily identified and can further be evaluated for developing an algorithm that can help in predictive maintenance of the available inventory.

No doubt, this is the future of the existing industries because the faults can be analyzed much before their occurrence. Even in some cases, almost one year before its occurrence. Due to which the industry can save millions by avoiding the downtime and their economic growth can be massive.

WSN APPLICATIONS

2.2.5.1 Life Cycle of Industrial IoT

The product lifecycle of industrial IoT is based on these four stages:

- Design: In this stage, the developer has to effectively integrate new features with old code without jeopardizing performance and security.
- Deployment: It can take several forms, including proof-of-concepts, pilots and commercial rollout.
- Management: To monitor the device's status, perform maintenance, issue updates and improve its performance, numerous stakeholders may need access.
- Decommissioning: End users and other stakeholders must be able to remove a device from service and onboard a new one in a timely and secure manner.

2.2.6 CYBER SECURITY

In Industry 4.0, each and every device is connected for real time data monitoring purposes. Therefore, the need for securing the manufacturing lines and the devices associated with it increases drastically [8]. As a result, industry requires secure and reliable connectivity as well as specific roles are required to be defined for each individual that is able to access the main server. Moreover, a fast connection plays a pivotal role in transmission of the manufacturing processes related equipment's data to the main server or cloud network.

Since I4.0 is a fully connected factory and involves different communication methods such as Wi-Fi 6, 5G, TSN Ethernet, etc. [8] end to end (E2E) connectivity becomes essential for industrial control systems (ICS). The recent increase in cyber-attack surface and SCADA systems has grown awareness of cyber risks and threats as well.

Cyber physical system (CPS) is a terminology applied to define human computer interface (HCI) [2]. Standalone connectivity of each and every device connected with the network differentiates the Industry 4.0 from other conventional industries [16].

In the manufacturing systems, systems like collaborative physical systems are used in dynamical reconfiguration of the CPS. It is mandatory to have a secure and established network connected to a cloud platform for the real time data transmission and monitoring [17].

Whereas, digital shadow of production reflects the entire production line of an industry in the simulation so that real time data can be correlated with the simulated data for making the predictive maintenance effective and for that a surety of a secure and healthy network is required which is only possible with the implementation of the cyber security algorithms.

Moreover, the use of proper sensors plays a crucial role in effectively identifying the industrial faults by using self-learning algorithms obtained using various deep learning algorithms. Each sensor's efficiency can be analyzed as they are standalone and are directly connected with the main server that could be a cloud network. This helps industry 4.0 to find the optimal utilization of individual devices that can increase the product rate which helps the industry to make huge profits.

2.2.6.1 Impact of Cyber Security

Securing the confidentiality, integrity and availability of data in a fully connected smart production environment in compulsory where billions of connected IoT devices are communicating and transmitting data. In such an automated and sophisticated environment if one device is tampered and manipulated it may affect the entire production life cycle.

2.2.7 CLOUD COMPUTING

For any industry, cloud-based models (public, private, hybrid, etc.) and infrastructures (SaaS, PaaS and IaaS) [18] serve as sound backgrounds for securing the connection and communication of each and every standalone device of the Industry 4.0's application portal [19,20]. Because of each standalone connectivity, the industry requires a fast and established network so that there would be no delays and the data could be processed with a high accuracy rate.

This satisfies the criteria of the real time processing. Furthermore, cloud networking is the backbone for disaster recovery (DR) so that precious industrial data could be secured and recovered in case of any failure of the physical server.

2.2.7.1 Life Cycle of Cloud Computing

In the current times, there has been an increased dependency on cloud-based applications and services. Majority of the supply chain breaches and cloud-vendor risks have also surfaced due to lack of appropriate security measures practiced at the vendor's end. As shown in Figure 2.4. In a fully connected factory where applications, services and real-time data is being analyzed in the cloud, the reliability, availability, scalability metrics are of utmost importance and if the required QoS (promised service level agreement) is not met it would lead the production environment susceptible to performance and security-based risks.

2.2.8   ADDITIVE MANUFACTURING

Additive manufacturing along with mass manufacturing goes side by side in the Industry 4.0. Industry 4.0 is based on customer feedback so it's obvious that small batches of customized products to be produced that might have complicated and light weight designs [14]. This will moreover also reduce various factors that might affect the cost factor of the developed product. As the product is much faster and cheaper than the conventional manufacturing due to the fused deposition method (FDM), sintering and laser cutting techniques. As customer demands are on the rise, so each industry plans in creating a separate production line for the additive manufacturing [15,21].

2.2.8.1   Impact on Product Life Cycle

Many significant mechanical and aerospace engineering firms have already included additive manufacturing into their long-term production plans. The main notion of a 3D printer is to integrate computer models of physical things and processes using e-manufacturing concepts. This adjustment necessitates a significant change in the corporate business model, which will influence either core or support activities.

2.2.9   AUGMENTED REALITY (AR)

AR plays a crucial role in a disaster or any catastrophic outcome of any industrial incident. It simulates those events in an actual scenario so that a real like virtual reality could be created [22]. This helps in training the worker specifically about the multiple faults that can occur in real time and their intensity. So that trained professionals could be trained enough for resolving any industrial faults that can occur in real time. It is a type of a simulation that is being played on an actual visual scenario [23,24].

2.2.9.1   Impact on Product Life Cycle

In most B2B stages, such as ideation, B2B product design visualization, simulation, prototyping, production, testing, training, and marketing, it is apparent that AR and VR are already making a difference or have a high potential.

2.3   CONCLUSION

We conclude this chapter by making a comparison between industry 4.0 and conventional industry. The table below summarises the key points of comparison.

Despite all of these distinctions, as seen in the chart above, Industry 4.0, unlike traditional industries, prioritises consumer or stakeholder feedback and machine-to-machine communication. The nine pillars of Industry 4.0 exist precisely for this reason.

Industry 4.0 refers to the total number of disruptive innovations produced and implemented across a value chain to meet the trends of digitization, transparency, mobility, modularization, network-collaboration and socialisation of goods and processes. Although most value chains are perceived as production-based, they may be

TABLE 2.1

Comparison between Legacy and Smart Manufacturing.

No.	Industry 4.0	Conventional industry
1.	In Industry 4.0 the resources are diverse therefore it is able to perform additive manufacturing and can work efficiently as it is able to have machine-to-machine communication and human computer Interface.	It has a fixed production line and has no room for additive manufacturing as well as due to lack of the machine-to-machine communication, tracking a fault in any existing industry is troublesome.
2.	The devices and the sensory units are able to reconfigure themselves using DHCP as their core networking protocol.	The devices are assigned static IPs therefore they are not able to reconfigure themselves to another switch if in any case a network failure occurs.
3.	Each and every device is interconnected with each other and with the main server. Moreover, they are also connected to cloud networking platforms in order to avoid any downtime due to IT failure.	Only few devices are connected to the main server therefore, it is hard in the conventional industry to identify and track fault in real time.
4.	As each and every device is connected with one another, therefore there is convergence in Industry 4.0. The tracking is much easier as the networking layers of various manufacturing processes are interconnected.	The network layers of the various manufacturing processes are separated therefore the tracking of the fault is complicated.
5.	There is a self-organization because of machine-to-machine communication, and they can compare the data with each other for its validation which makes them self-learn any fault and can perform predictive maintenance according to it.	The machines are preprogrammed and they lack in machine-to-machine communication due to which the machines are unable to self-learn.
6.	There is the use of big data and deep learning as there is a massive amount of data that is acquired from various manufacturing processes.	Each manufacturing process is able to record isolated information of any manufacturing process.

augmented with logistical operations [22], Industry 4.0 penetrates the whole value chain of the organisation and their future [22].

Moreover, these concepts of Industry 4.0 are already implemented and tested in developed countries like Germany, Israel, and the United Kingdom and have proved their mettle from time in and time out. Therefore, it will a great opportunity for the developing countries like Pakistan, Bangladesh and India to make their industries affirm these concepts and adapt them as the future of the Industrial revolution is already in the works in form of Industry 5.0. Which is a modified version of Industry 4.0 and aims at working side by side with the humans to enhance industrial productivity. Industrial automation is significant and European Economic Social Committee (EESC) has called its proliferation inevitable [25].

REFERENCES

[1] Y. Lu, "Industry 4.0: A survey on technologies, applications and open research issues," *J. Ind. Inf. Integr.*, vol. 6, pp. 1–10, 2017.

[2] S. Vaidya, P. Ambad, and S. Bhosle, "Industry 4.0—a glimpse," *Procedia Manuf.*, vol. 20, pp. 233–238, 2018.

[3] S. Bag, S. Gupta, and S. Kumar, "Industry 4.0 adoption and 10R advance manufacturing capabilities for sustainable development," *Int. J. Prod. Econ.*, vol. 231, p. 107844, 2021.

[4] M. Elsisi, M.-Q. Tran, K. Mahmoud, M. Lehtonen, and M. M. F. Darwish, "Deep learning-based Industry 4.0 and internet of things towards effective energy management for smart buildings," *Sensors*, vol. 21, no. 4, p. 1038, 2021.

[5] "From mass production to mass customization," 2012. https://thesismusen2012.wordpress.com/2012/10/09/presentation-section3-from-mass-production-to-mass-customization/ (accessed May 29, 2022).

[6] D. Ivanov, C. S. Tang, A. Dolgui, D. Battini, and A. Das, "Researchers' perspectives on Industry 4.0: Multi-disciplinary analysis and opportunities for operations management," *Int. J. Prod. Res.*, vol. 59, no. 7, pp. 2055–2078, 2021.

[7] W. P. Neumann, S. Winkelhaus, E. H. Grosse, and C. H. Glock, "Industry 4.0 and the human factor—A systems framework and analysis methodology for successful development," *Int. J. Prod. Econ.*, vol. 233, p. 107992, 2021.

[8] "Unix Packaging investing $19 million to build new plant in Burke County," 2020. www.ncconstructionnews.com/unix-packaging-investing-19m-to-build-new-plant-in-burke-county/ (accessed May 29, 2022).

[9] L. L. Dhirani and T. Newe, "Hybrid cloud SLAs for Industry 4.0: bridging the gap," *Ann. Emerg. Technol. Comput. (AETiC)*, Print ISSN, pp. 281–2516, 2020.

[10] "Industry 4.0—A realistic pathway to the smart factory." https://slcontrols.com/en/industry-4-0-realistic-pathway-smart-factory/.

[11] L. L. Dhirani, E. Armstrong, and T. Newe, "Industrial IoT, cyber threats, and standards landscape: Evaluation and roadmap," *Sensors*, vol. 21, no. 11, p. 3901, 2021.

[12] "The dawn of the smart factory," 2013. www.industryweek.com/technology-and-iiot/article/21959512/the-dawn-of-the-smart-factory (accessed May 29, 2022).

[13] M. C. & Robotics, "No title," [Online]. https://motioncontrolsrobotics.com/robot-life-cycle-faqs/.

[14] "Industry 4.0 and smart factory," www.slideshare.net/AlaaKhamis/industry-40-and-smart-factory (accessed May 29, 2022).

[15] "Applications of industrial IoT," www.sketchbubble.com/en/presentation-industrial-iot.html.

[16] C. Cimini, R. Pinto, G. Pezzotta, and P. Gaiardelli, "The transition towards Industry 4.0: Business opportunities and expected impacts for suppliers and manufacturers," in *APMS (IFIP international conference on advances in production management systems). IFIP Advances in Information and Communication Technology*, vol. 513. New York: Springer, 2017, pp. 119–126.

[17] "Gaps in Industry 4.0 readiness contribute to Industrie 4.0," [Online]. www.pinterest.com/pin/781444972819722217/?amp_client_id=CLIENT_ID(_)&mweb_unauth_id=%7B%7Bdefault.session%7D%7D&simplified=true.

[18] L. L. Dhirani, T. Newe, and S. Nizamani, "Tenant-vendor and third-party agreements for the cloud: Considerations for security provision," *Int. J. Softw. Eng. Its Appl.*, vol. 10, no. 12, pp. 449–460, 2016.

[19] L. L. Dhirani, T. Newe, E. Lewis, and S. Nizamani, "Cloud computing and Internet of things fusion: Cost issues," in *2017 eleventh international conference on sensing technology (ICST)*. Piscataway, NJ: IEEE, 2017, pp. 1–6.

[20] A. Dey, "Internet of things (IoT)—security, privacy, applications & trends," https://medium.com/@arindey/internet-of-things-iot-security-privacy-applications-trends-3708953c6200.

[21] "Science, Technology, Engineering, and Math (STEM) overview | career cluster/industry video series," www.youtube.com/watch?v=9ZdNopKi7M0.

[22] G. M. Santi, A. Ceruti, A. Liverani, and F. Osti, "Augmented reality in Industry 4.0 and future innovation programs," *Technologies*, vol. 9, no. 2, p. 33, 2021.

[23] E. Marino, L. Barbieri, B. Colacino, A. K. Fleri, and F. Bruno, "An Augmented reality inspection tool to support workers in Industry 4.0 environments," *Comput. Ind.*, vol. 127, p. 103412, 2021.

[24] Y. El Filali and S. Krit, "Augmented reality types and popular use cases," *Int. J. Eng. Sci. Math.*, vol. 8, no. 4, pp. 91–97, 2019.

[25] J. Jardine, "Industry 5.0: Top 3 things you need to know," 2020. www.mastercontrol.com/gxp-lifeline/3-things-you-need-to-know-about-industry-5.0/.

3 Security at the Internet of Things

*Yousaf Ali, Syed Waqar Shah
and Wasim A. Khan*

CONTENTS

DOI: 10.1201/9781003220985-3

Industry 4.0 or Industrial Revolution 4.0 is the current data exchange and automation trend in manufacturing and machine tools technology. This includes the use of internet of things (IoT), cloud computing artificial intelligence and cyber physical system that gives birth to smart manufacturing. The smart industries are equipped with sensing system, the powerful microcontrollers, the embedded software and the automated actuator for making smart data collection and better decision making. Smart industries also add to the productivity and enhanced quality of product by replacing manual quality inspection model with highly sophisticated and faster quality inspection system that saves time and money.

The data and communication system security issues are not considered yet while designing the industrial system because the network of the industry was not accessible from outside and it was considered as isolated system. But with the advent of IoT the devices and hence the industrial machining tools are connected to the external world. This interconnection of industries in Industry 4.0 has given rise to the important of data and network security in industrial IoT and hence taking the right steps for security in IoT are priority while designing industrial communication network. We have therefore discussed the most common vulnerabilities in the security of IoT enable machining. The security threats at each layer of industrial IoT are mentioned and the remedies are suggested to help improve the security of IoT enable machining.

3.1   INTRODUCTION

The IoT was first proposed by Ashton et al. at Massachusetts Institute of Technology [1]. With the advent of 4G technology and now the 5G technology, the users of IoT are increasing very rapidly, and it is expected that the number of devices connected via IoT across the world will exceed one hundred billion by the year 2022 [2]. IoT is the interconnection of devices, apparatus and appliances at home, industries, power houses and educational institutes through the internet. The devices are controlled remotely via internet with the help of sensors. IoT is changing life of humans and making it more comfortable. IoT communication takes place with a set of predefined model and protocol. Security in the IoT is of great importance because of wide variety of devices, sensor and computers used in the environment. The IoT communication takes place on four layers as depicted in the Figure 3.1. Although these four layers are not an adopted standard but in most of the literature it is used, so we will be using this model for our study also. All the four layers are detailed in the following paragraphs.

3.1.1   PERCEPTUAL LAYER

Perceptual layer is equivalent to physical layer of OSI reference model. Perceptual layer of IoT stack consists of different devices including actuators, sensors, radio frequency ID (RFIDs) and other physical devices, which sense the real environment,

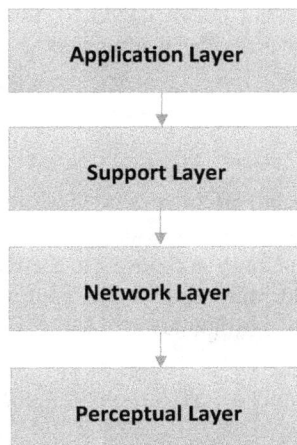

FIGURE 3.1   IoT Layers.

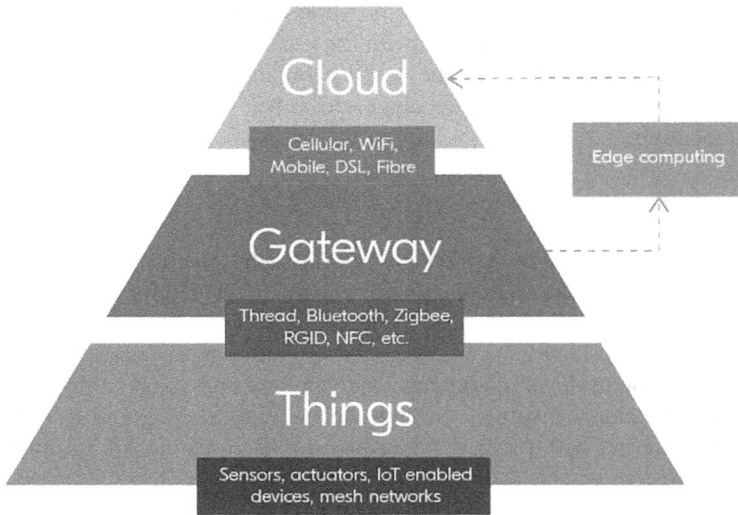

FIGURE 3.2   IoT Architecture (www.blog.nordicsemi.com).

such as temperature, weather, pressure, light, wind speed, water level, and operations in agriculture, industries, institutes and homes. These devices collect data and send it to internet gateways.

3.1.2 NETWORK LAYER

Networks layer of IoT is responsible for assigning IP addresses to network nodes, it is responsible to authorize the access to network. Routing, route discovery, route optimization and route maintenance take place at this layer. This layer may use different technologies such as mobile ad hoc network (MANETs), Wi-Fi technology and satellite communication. The devices at this layer may use different routing protocols like OSPF, ISIS, AODV and DSDV.

3.1.3 SUPPORT LAYER

Support layer gives a feasible and effective ground for IoT users. Different IoT applications can use the cloud, and can be accessed through internet by the resource contained devices. It also gives memory and power to the resource nodes. Technologies used at this layer include cloud computing, fog computing and intelligent computing. This layer is also named as recognition layer of IoT.

3.1.4 APPLICATION LAYER

This layer provides the access to user, engineers and technicians at different industries. The application may include smart home, smart institutes, smart grids and smart agricultures.

3.2 APPLICATIONS OF IOT

Applications of IoT include but not limited to smart homes, smart grids, smart class-rooms, wearables, smart cities, energy management, utility management and farming. A few of the aforementioned are further explained below.

3.2.1 SMART HOMES

With IoT in picture we can switch on our air conditioner before reaching our home from office, we can switch on and off all our home lights, Fans and other appliances remotely using internet. Smart home products are promised to save our time, energy and money.

3.2.2 SMART INDUSTRIES

Application of IoT in industry has given birth to Industry 4.0 or industrial IoT. It is a new trend in industrial sector. It is empowering industries with wireless sensors and actuators and big data analytics to create more efficient machines. The driving force behind industrial IoT is that IoT enabled machines are more accurate than humans in communication. And this data can help industries pick troubles and faults sooner.

IoT also provide good quality control and sustainability. IoT is efficiently used for tracking goods and supply chain management. GE predicted that use of IoT may generate $10 trillion to $15 trillion in GDP in the whole world over the next two decades.

3.2.3 IoT IN FARMING

Population of the world has increased many folds in recent years and the demand for food is also increased. States are encouraging farmers to use new technology and research to grow food production. Use of IoT in farming make it smart farming. Sensors are deployed in fields to sense moisture condition, water level and growth of plants. The use of fertilizers and the time is also decided based on IoT devices used.

3.2.4 ENERGY MANAGEMENT

Smart grids allow the use of IoT in substation and power plants to analyze the behavior of plant machinery and to predict the energy consumption and production. Power grids in future will be smart and highly reliable smart grid also allow us to analyze health of plant equipment. Smart grids will also be able to find sources of energy shortage more quickly and at individual levels like the solar panel, and thus makes distributed power system possible.

3.2.5 IoT IN HEALTHCARE (WBANs)

IoT have got many applications in the healthcare. These applications have got a lot of attention from researcher around the globe new research are carried out and new methods are developed to make the patient fight with the diseases more effectively.

The IoT application in healthcare has made the processes more accurate and easier. WBANs the wireless body networks is one of application of the IoT is making the patient to wear some sensors and devices and they are monitored by doctors remotely. The doctors collect the data the sensor and diagnose the patient accordingly.

3.2.6   Sensors in IoT Application

Sensors are vital component of IoT architecture. They are part of perceptual layer of IoT. The figure below show position of sensors in IoT architecture [3]. Sensors are used to detect and measure different physical quantities in IoT application. Senor collects the data and send it to the IoT gateway for decision making and performing the respective action in different industrial application. Some of the very common sensors used are discussed in the following paragraphs.

3.2.6.1   Sensors in Smart Homes

Smart home makes the use of many sensors. Some common sensors include fire detection sensors and moisture sensors to detect leakage in drainage system. Other sensors are temperature sensors to monitor room temperature and automate room air conditioning. Motion sensors are used in home security application which detect if there is any motion in the home and act as guard when you are not at home. Infrared sensors are proximity sensors which detect whether the doors are open or closed.

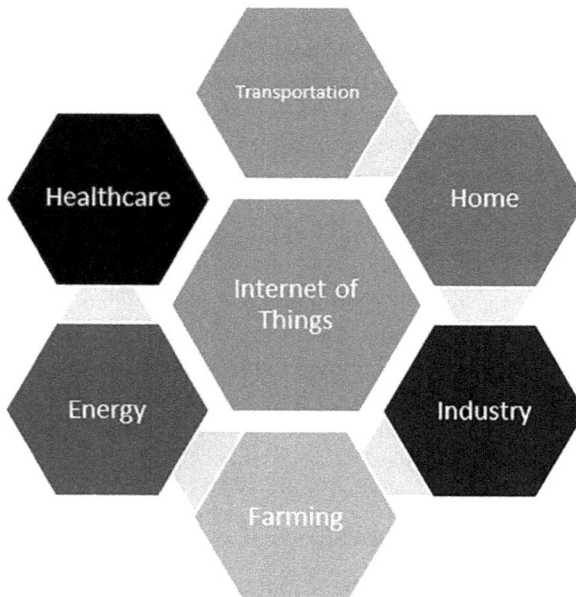

FIGURE 3.3   IoT Applications.

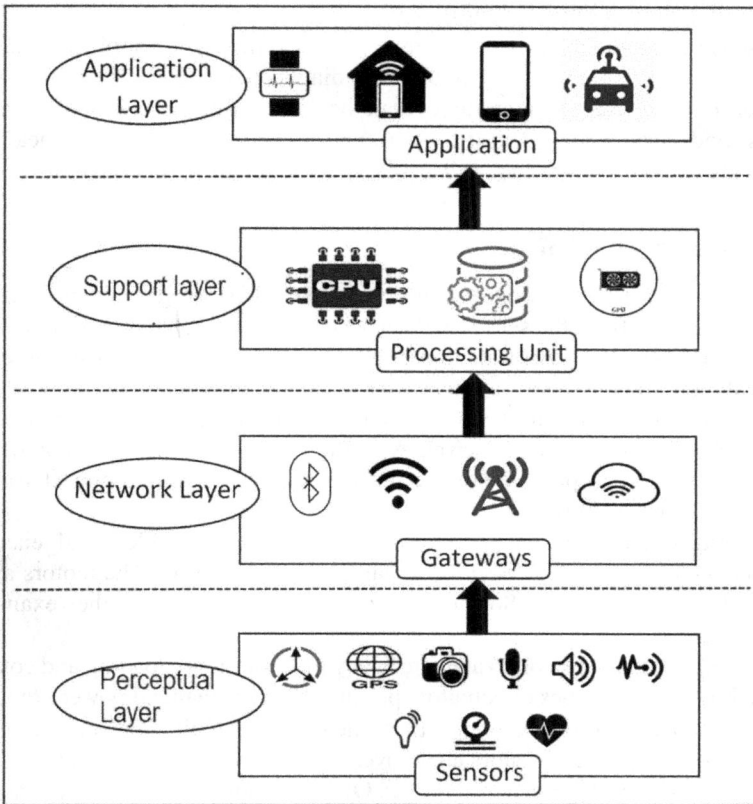

FIGURE 3.4   Sensors in IoT Architecture.

3.2.6.2   Sensors in Smart Industries

Pressure sensors are most widely used sensor in industries, pressure sensor detect and measure the pressure of liquid and gas and activate actuators like motors. Proximity sensor and limit switches are used to detect contact between different parts of machines. The other sensors used in industry includes level sensors and gyroscopes. Gyroscopes are used to measure the angular motion in different industrial applications.

3.2.6.3   Sensors in Agriculture Sector

The IoT sensors used in agriculture mainly collect the environment conditions like temperature moisture and humidity. Other sensors used in agriculture are used to find the machinery metrics. The moisture sensors are used to collect information about soil quality, leaf condition and overall growth of crops. The sensors are helping the farmers to automate the farming cycles including irrigation, pest control and fertilizing.

3.2.6.4 Sensors in Smart Healthcare

Image sensors are widely used sensor in IoT enabled healthcare applications. Image sensors are used in cardiovascular diseases diagnostics and monitoring. Wireless body sensors are used to monitor overall health condition of patients remotely. Wireless body sensor can monitor temperature, blood pressure and heartbeat for all day and also help in prevention of diseases.

3.2.7 ACTUATORS IN IoT

Actuators are the machines or system components that run the system. Actuators receive the signals from the sensor and act according to the data received, for example a cooling fan in the power transformer is an actuator which starts when it receives the signal from the temperature sensors that the temperature has risen above a predefined value. Similarly, the level sensor may activate the water pump when the water level is below a specified level. An actuator can be a servo motor which is activated in a direction provided by the sensor data. The following flow chart shows how the actuators are activated.

The actuators can be electrical actuator which coverts electrical energy to mechanical energy, they are used for linear or angular motion. The motors are one type of electrical actuators. Solenoid based electronic bell is another example of electrical actuators.

Hydraulic actuators use oil to activate heavy machineries in loading and construction machines. These types of actuators produce high mechanical power. Pneumatic actuators are used to produce energy from the pressure of air. The vacuum or compression system in these actuators are converted to linear or rotary motions. These actuators are also used in industrial robots. Other two common types of actuators are magnetic actuators and mechanical actuators.

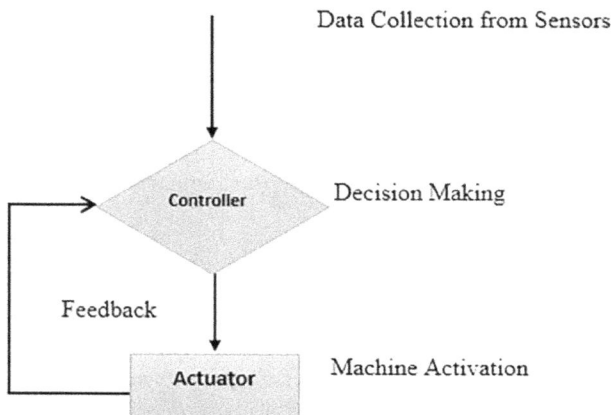

FIGURE 3.5   Actuator in IoT.

3.3 SECURITY IN NETWORKS

Computer security or cyber security is the prevention of computer data from theft, damage, illegal access and defacing. This data can be bank account, credit card details, passwords and other confidential data. Security can be physical security restricting the un-authorizes user to use the computer and communication devices, Security can be software based like antivirus software to avoid damage of data or to prevent denial of service attacks. Network security is based on the following four principles.

3.3.1 AUTHENTICATION

Authentication is process to make sure the identity of user is obtained, and the user is the list of allowed and legitimate user. An authentication server is required in the network architecture to make the authentication possible. The user must be made sure they are communicating with whom they are intended to communicate.

3.3.2 AVAILABILITY

Data should be available to use when required. Data availability is made sure by providing high performance server and providing redundant links or stacked devices. In case the service is accessed by many users, the service is made available by providing multiple servers.

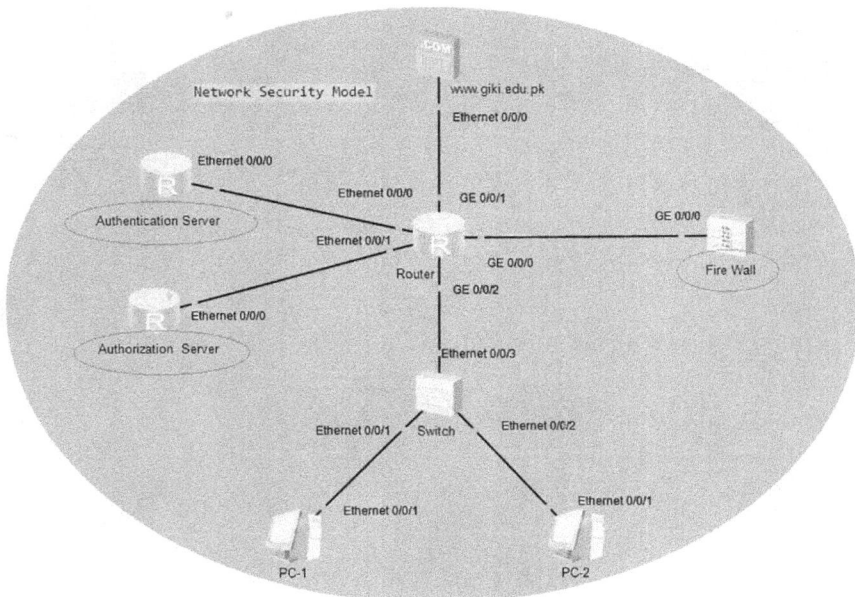

FIGURE 3.6 Network Security Model.

3.3.3 Authorization

Only legal and authorized persons will be allowed to access and use data. Authorization comes after authentication. Once the user is authenticated, the system checks whether to allow the user to use the service or not. A separate authorization server is required for authorizing users.

3.3.4 Integrity

The data on servers can only be modified by legitimate user or administrator(s). Special security mechanism like encrypted password access and multi-step verification may be required while accessing the server for data modification.

Security model used in the mid-level enterprises networks are depicted in the figure below. The firewall shown in the figure below provides the mechanism to monitor the incoming and outgoing data to and from the secured network.

3.4 SECURITY ATTACKS ON THE INTERNET OF THINGS

Security at each four layers is discussed below in details, the aim of security is to make the data transfer secure and allow authenticated access to network nodes

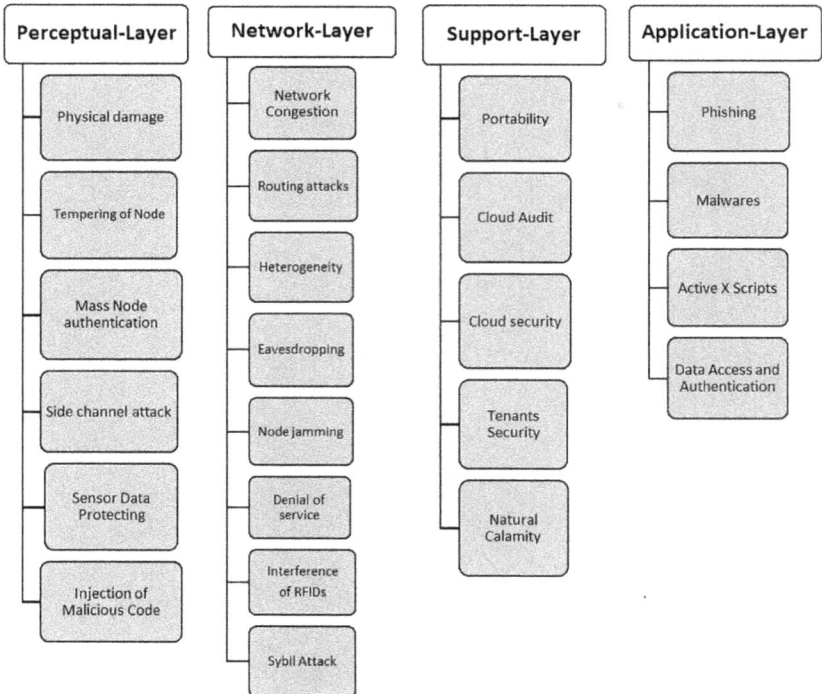

FIGURE 3.7 IoT Security Attacks.

and gateway. The figure below provides the list of security attacks in each four layers of IoT.

3.4.1 SECURITY AT PERCEPTUAL LAYER OF IoT

Since perceptual layer is having physical devices like sensors, Bluetooth, radio frequency IDs and other devices they are more vulnerable to attacks. These IoT physical device are installed in open areas in different application that's why they face many attacks, some are explained below.

3.4.1.1 Physical Damage

Since the nodes and sensors in the IoT networks are open and easily accessible, The attacker may damage the devices physically, making the service unavailable for users. The nodes installed in open fields are more vulnerable to attacks. The IoT device may also get physically damaged if they are operated inadequately. IoT enable devices may stop in the middle of operation if they are accessed by unauthorized personals. The IoT connected vehicle for example may be turned off in the journey or their brakes might be disabled by the intruders.

3.4.1.2 Tempering of Nodes

Node tempering is the physical damage to the IoT devices at perceptual layer. Node tempering may be in the form of changes in hardware of a node or changing one node with another node. By changing nodes or changing hardware like memory from a node can provide illegal access to the attackers [4]. This information may be very sensitive like passwords, keys, bank accounts details or even the routing table information of the healthy network. In these attacks the attacker has physical access to the network devices in IoT communication system.

3.4.1.3 Mass Node Authentication

If the IoT communication network consist of a large number of nodes and sensor devices. It will be required to have authentication request and a large number of authentication request may badly affect the performance of IoT network. Increasing the end-to-end delay, or the time taken by packet from source to destination.

3.4.1.4 Side Channel Attack

In side channel attack the attacker get access to the channel information of a network by physically accessing the node. The channel information may be the time delay, i.e., time consumption, the power requirement of nodes and the sensors properties to attack the cryptographic mechanism of the network [5].

3.4.1.5 Sensor Data Protecting

Since the sensors are placed in open area so the attacker may place another sensor in the IoT environment to sense the data and get the values of environment. However, the authenticity and integrity are of great importance and thus it must be make sure they are secured.

3.4.1.6 Injection of Malicious Code

Malicious code injection is the insertion of the malicious code and other executable files in to the healthy node [6]. This insertion may provide the attacker with the illegal access to the network and thereby stealing information. This attack may also destroy the information files in the healthy node.

3.4.2 SECURITY AT NETWORK LAYER OF IoT

The network layer attack consists of denial-of-service attacks, eavesdropping attack. Black hole attack, wormhole attack and other man-in-the-middle attacks. The IoT core network are secured against many network layer attacks but still there are chances of malicious activities at network layer. These activities affect the integrity, authenticity and availability of the healthy node of the network

3.4.2.1 Network Congestion Problems

Network congestion takes place when the IoT network consist of a large number of nodes. Because of large number of nodes, the big amount of data transfer takes place thereby making the network overloaded. Secondly the large amount of authentication may also cause network to be congested. This problem can be solved by installing efficient network layer devices having feasible authentication and data transfer mechanism and competent transport protocols.

3.4.2.2 Routing Attacks

In this attack the attacker get access to the routing information of the network and get route to different devices, get the subnet masks and IP addresses of different nodes. The attacker may access the routing table and make changes in the table and forward it to the other nodes to create loops in the routing. The attacker may also broadcast false route messages to the neighbor nodes and thus packet drop, and information loss takes place.

3.4.2.3 Heterogeneity Problem

IoT uses many technologies at network layer of the model so making the IoT system wider and more complex in term of access. As the number of devices connected across the IoT, this increases the heterogeneity of IoT increases because some of the devices may use Ethernet, Zigbee, Wi-Fi and Bluetooth at network layer. Similarly, the heterogeneous type of data may require different level of quality-of-service requirement. Because of complex access methods the security and integrity of the network becomes more difficult.

3.4.2.4 Eavesdropping Attack

These attacks use the sniffing tools like packet sniffing to steal the data. Since the sensors are of a wireless nature so they are more east to be sniffed. It is the sniffing of data in the Bluetooth, infrared, Wi-Fi, wireless sensor network and other wireless technologies because of vicinity.

3.4.2.5   Node Jamming in WSN

The node jamming attack is same as that of radio frequency attacks. The attackers here also get interference in the radio frequency of information signal of sensor and causes the denial of access to the Service. WSN is one of the forms of the attack where radio frequency signal is noised [7]. This is one of the type of DoS attacks.

3.4.2.6   Denial of Service

The attacker sends a large amount of data to the network more above the capacity of the network to handle the data and thus the network is not available for the useful services to legal users. These attacks include black hole attack and wormhole attacks.

3.4.2.7   Interference of RFIDs

RFID devices works of the radio frequencies so in these attacks the attackers add noise to the radio waves signal of IoT devices and therefore destroying or changing the information signals. This changing or destroying of signal can causes the DoS denial of Service attack at network layer of IoT.

3.4.2.8   Sybil Attack

In this type of attack only a single node called sybil attack node broadcast its identity as of other many nodes and thus disguising the network nodes [8]. This sybil attack node can harm the network in many ways like sending false route reply to messages and disrupt the routing process. This attack is considered to be one of the biggest issues when nodes are connected in point-to-point mode. This attack is targeting the computational power and power consumption of mobile nodes. This kind of attacks poses serious threat to routing mechanism in WBANs [9].

3.4.2.9   Spoofing of RFID

In this kind of attack, the malicious node reads the RFID tags a healthy node and therefore using these tags it sends false information to other nodes and get access to the network resources.

3.4.3   Security at Support Layer of IoT

Support layer is the third layer of IoT communication stack. The support layer is responsible to make sure the data is sent by the legal and authentic source and is also responsible to secure the data from threats. Cloud Security Alliance (CSA) has made many security standards to secure the data from malicious users. CSA has made (SCAP) [10] and (TCG) [11]. The SCAP is Security Content Automation Protocol used for listing the security issues in the communication stack of IoT. TCG is Trusted Computing Group responsible for protecting the device and maintaining against malware. Support layer security issues are discussed below.

3.4.3.1 Portability

Data portability is the protection of data from other users and allowing the owner to use the stored data in cloud for other application by the original developer. Since there are many cloud service providers and each uses the propriety protocols, when the user migrates from one service provider to the other it creates problems. Security vulnerability is also produced by this heterogeneity [12].

3.4.3.2 Audit of Cloud

The cloud audit is concerned about how the cloud service provider addresses and control the framework. The CSA security alliance set several standards for cloud service providers. It is necessary for building user trust to continuously make audits to check the cloud service provider comply with the security standards.

3.4.3.3 Security at Cloud

The data in IoT is stored in data cloud, So the data need to be secured from attacker and malicious user. Special arrangements shall be made to monitor the data transfer from and to the data cloud. Certain applications and tools can help in protecting data loss. Other techniques like dispersion and segmentation can be further used to secure data in the data cloud [13].

3.4.3.4 Security of Tenants Data

The data of many users are normally stored in the same physical drives in the data cloud such users are named as tenant's user.one of the tenant can steal the data of his tenant since the data of both are placed at the same physical location.

3.4.3.5 Continuous Service in Natural Calamity

Data cloud service provider must make sure that the service is available in case of natural calamities earthquake, fire, floods and other natural disasters. The physical location of the data servers should be such that there are minimal chances of natural disasters. Secondly it must be easily accessible to the emergency response like fire brigade and other disaster management authorities. Thirdly the data cloud service must have some data recovery and back up mechanism [14].

3.4.3.6 Virtualization Security

Security virtualization is the shift of security protocol from one hardware device to software that can be easily moved between hardware or run in the data cloud. Different cloud service provider is using different virtualization techniques. The machine some time bypass the security protocol and thus making it difficult to audit data cloud [15].

3.4.4 SECURITY AT APPLICATION LAYER OF IoT

Since the applications of IoT are many, so they used different many devices and software for running the network. There do not exist and standard application. However, data sharing is one of the characteristics of IoT application layer. Sharing of data are vulnerable to malicious users. Application layer are security issues and are discussed below:

3.4.4.1 Phishing Attacks

In this attack the disguised emailed are used to steal user data. This attack makes the receiver of email believed that this sent information is what is needed to him and therefore steal username, password, and other encrypted data [16].

3.4.4.2 Malwares Attack

Malwares are the software design to attack the cyber devices. Malware attackers may use the codes to steal user data, corrupt the data or may cause the DoS denial of service attacks. Some of the malwares are worms, Trojan horses and other viruses which are used by attackers to manipulate user data.

3.4.4.3 The Active X Scripts

The attacker can inject active x scripts in to the IoT network node via internet and thus making the IoT user make to run their active x script and thus making all the network prone to the attacks [17]. These scripts cause the system to close down and sometime by running this script the data get stolen. These attacked are targeting the data integrity of IoT system [18].

3.4.4.4 Data Access and Authentication

An IoT-based system has many users and they are using different level of privileges, some has access to only a few types of data while other can access all the available information. The access control and authentication should be applied to protect network from the attacks of application layer.

3.5 SECURITY REQUIREMENTS IN IOT

The following paragraphs provide how the data at each layer can be secured.

3.5.1 SECURITY REQUIREMENTS OF PERCEPTUAL LAYER OF IoT

The physical devices, the nodes, the sensors and other network devices of IoT network must be physically secured from illegal access. They must be placed in the vicinity of other IoT device and proper physical security be provided. Only authorized persons should be allowed to access the sensors physically. The availability, authenticity and integrity of data must be made sure against attackers, for this purpose different encryption techniques and algorithms must be designed. Key management is also an issue to be addressed in IoT.

3.5.2 SECURITY REQUIREMENTS OF NETWORK LAYER OF IoT

Much has been done to secure the network against the network layer attack and existing mechanism are enough to tackle network layer security issue but still some security risks exist and they are more dangerous in the IoT, like DoS and D-DoS attacks, denial of service and distributed denial of service must be stopped at this layer. The dynamic protocols used for routing and optimization at network layer should be smart enough to take care of routing attacks, congestions problems, end to end delay and other issues. The protocols at this layer must be modified after specific intervals.

3.5.3   SECURITY REQUIREMENTS OF SUPPORT LAYER OF IoT

The data of IoT user are on data clouds. So, these data clouds must be secured so that the user data are not accessed by attacker nodes. As discussed, earlier CSA and TCG are responsible for security of data at clouds. CSA must make sure the compliance of the security standards must be continuously checked and IoT systems must only use those clouds services which do comply with the security standards provided by the CSA. Also, continuous cloud audit must take place to make sure the data is secured. Virtualization techniques must also make sure the security protocols are not bypassed.

3.5.4   SECURITY REQUIREMENTS OF APPLICATION LAYER OF IoT

Application layers attacks can be stopped by using strong encryption and encoding techniques, difficult to guess authentication password and other access control

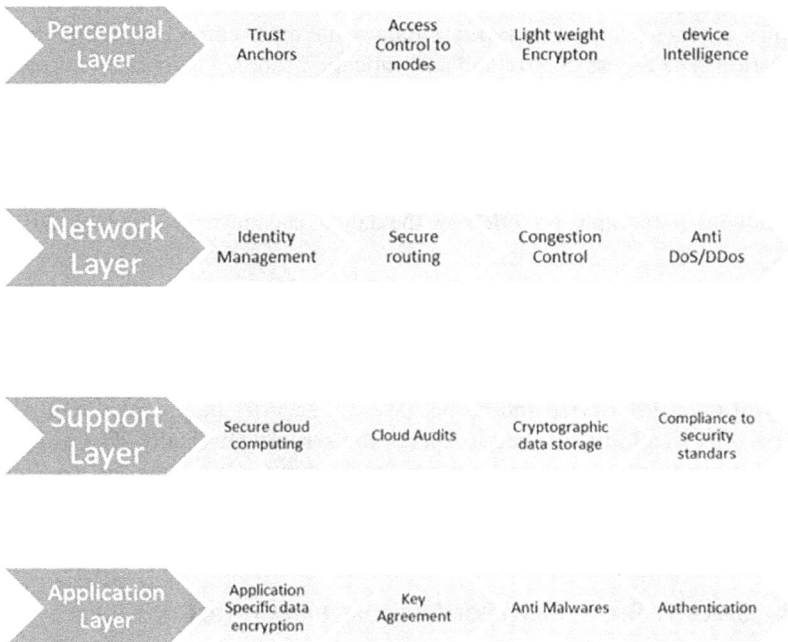

FIGURE 3.8   Security Requirements in IoT.

mechanism. The nodes must also be secured from malware attacks by using antivirus software. The IoT users must be aware of security attacks and loss in case of these attacks. The data generated at application layer must be end to end encrypted with more secured encryption techniques. User privacy is also a requirement of security at application layer of IoT.

3.6 CONCLUSION

Security is one of the key aspects of any network, same is the case with IoT networks. With evolution of 5G communication it is the hot research topic to secure the IoT-based network from different attacks. We have discussed the IoT architecture and a four-layer model of IoT in this chapter in detail. The applications of IoT are discussed. The sensors and actuator used for different application are mentioned. In this work we have provided an overview of security attack in the IoT networks. Several suggestions have been provided to secure IoT network against different types of attack at four-layer of IoT communication stack.

REFERENCES

[1] Ashton, K., 2009. That 'internet of things' thing. *RFID Journal, 22*(7), pp. 97–114.
[2] Sundmaeker, H., Guillemin, P., Friess, P. and Woelfflé, S., 2010. Vision and challenges for realising the Internet of things. *Cluster of European Research Projects on the Internet of Things, European Commision, 3*(3), pp. 34–36.
[3] Sikder, A.K., Petracca, G., Aksu, H., Jaeger, T. and Uluagac, A.S., 2018. A survey on sensor-based threats to internet-of-things (iot) devices and applications. *arXiv preprint arXiv:1802.02041.*
[4] Kaminow, I.P. and Li, T., eds., 2002. *Optical fiber telecommunications IV B: Systems and impairments* (Vol. 2). Elsevier.
[5] Perrig, A., Stankovic, J. and Wagner, D., 2004. Security in wireless sensor networks. *Communications of the ACM, 47*(6), pp. 53–57.
[6] Zhao, K. and Ge, L., 2013, December. A survey on the internet of things security. In *2013 Ninth international conference on computational intelligence and security* (pp. 663–667). IEEE.
[7] Francillon, A. and Castelluccia, C., 2008, October. Code injection attacks on harvard-architecture devices. In *Proceedings of the 15th ACM conference on computer and communications security* (pp. 15–26). SIGSAC.
[8] Mpitziopoulos, A., Gavalas, D., Konstantopoulos, C. and Pantziou, G., 2009. A survey on jamming attacks and countermeasures in WSNs. *IEEE Communications Surveys & Tutorials, 11*(4), pp. 42–56.
[9] Gothwal, R., Tiwari, S., Shivani, S. and Khurana, M., 2020. Progressive advancements in security challenges, issues, and solutions in e-health systems. In *Intelligent data security solutions for e-health applications* (pp. 237–254). Academic Press.
[10] Newsome, J., Shi, E., Song, D. and Perrig, A., 2004, April. The sybil attack in sensor networks: analysis & defenses. In *Third international symposium on information processing in sensor networks, 2004. IPSN 2004* (pp. 259–268). IEEE.
[11] Open SCAP, http://open-scap.org/page/Main_Page [Online; accessed 25.01.14].
[12] Open-source TCG software stack in C, http://trousers sourceforge.net/; 2011 [Online; accessed 25.01.14].

[13] The treacherous 12, Cloud computing top threats 2016, Top threats working group, Cloud Security Alliance (CSA), https://library.cyentia.com/report/report_001352.html; 2017 [Online; 27.11.21].

[14] Heer, T., Garcia-Morchon, O., Hummen, R., Keoh, S.L., Kumar, S.S. and Wehrle, K., 2011. Security challenges in the IP-based Internet of things. *Wireless Personal Communications*, *61*(3), pp. 527–542.

[15] Jagatic, T.N., Johnson, N.A., Jakobsson, M. and Menczer, F., 2007. Social phishing. *Communications of the ACM*, *50*(10), pp. 94–100.

[16] Kim, C., Talipov, E. and Ahn, B., 2006, August. A reverse AODV routing protocol in ad hoc mobile networks. In *International conference on embedded and ubiquitous computing* (pp. 522–531). Springer.

[17] Zarei, M., 2009, February. Reverse AODV routing protocol extension using learning Automata in ad hoc networks. In *2009 2nd international conference on computer, control and communication* (pp. 1–5). IEEE.

[18] Kanagachidambaresan, G.R., Anand, R., Balasubramanian, E. and Mahima, V., eds., 2020. *Internet of things for Industry 4.0: Design, challenges and solutions*. Springer.

4 Internet of Things Advancements and Challenges in Manufacturing Industries

*S. S. Farooq, Muhammad F. Shah,
Muhammad S. Sahar, Ghias M. Khan,
Zareena Kausar and Muhammad U. Farooq*

CONTENTS

This work relates to the 'Industry 4.0 Perspective' theme of WRE 2021. It highlights the design and development of IoT based 3D printer according to the requirements of machine under consideration. The term 'internet of things (IoT)' describes the network of physical objects that are embedded with sensors, software, and other technologies that is used for the purpose of connecting and exchanging data with other devices and systems over the internet. This work incorporates the concept of internet of things in industry by designing a 3D printer workstation that is IoT enabled for remote use of printer; one that is designed around principles and a framework that enables it to be industry scalable and serve as a demonstration of concept for total IoT enabled industry under the banner of Industry 4.0. The ultimate vision of this project is to establish inter machine communication for optimized industrial manufacturing and quality control, but the implementation of this concept is beyond the scope of this study.

DOI: 10.1201/9781003220985-4

4.1   INTRODUCTION

With the advent of technology, many manufacturing industries all over the world have shifted from conventional methods to smart and digital methods. With the growing product demands it sometimes become very difficult for manufacturing industries to maintain production targets with conventional methods. Aspects of the Industry 4.0 technological revolution encompass technologies used in embedded production systems along with intelligent production methods. This leads to the introduction of a new technological age that changes the entirety of the production value chains, business models, and industry value chains [1]. A computerized monitoring and control system is being used with the aid of sensors, information technology and other software. These are used to interlink the data with systems and devices with the help of the internet to exchange information. This concept of internet of things (IoT) is used in many fields of manufacturing engineering such as HVAC industry, automation and control industry, medical industry, and many more. The IoT is defined as a pattern where the different objects in a system are interlinked using the specified sensors, actuators, and processors. The former two elements are used to sense the surrounding information, collect, and interpret it, and transfer it to the IoT gateways. The latter is used to convert the unprocessed data into a digital form that is readable and understandable.

The three types of data that go into manufacturing are device, product, and command data. A bridge for communication between entities and information in intelligent manufacturing, the industrial internet of things facilitates data exchange [2]. When and where they are needed across a holistic manufacturing supply chain, multiple industries, small and medium-sized enterprises, and large corporations, it enables all physical processes and information flow to be accessible [3]. IoT has pushed the manufacturing sector to go for sustainable solutions and better management strategies and solutions for a complex framework of machines. As a result, due to market competition, the production rate has been increased.

Manufacturing paradigms shift frequently, and support for new manufacturing paradigms is commonly brought about by advancements in information technology (IT). CNC and industrial robots were widely adopted, which made flexible manufacturing systems (FMSs) possible. Technologies for computer aided drawing (CAD), computer aided manufacturing (CAM), and computer aided process planning (CAPP) made computer integrated manufacturing (CIM) feasible. Companies increasingly utilize IT service providers to augment or replace their conventional systems [4].

This chapter provides an overview of IoT, its usage and challenges in manufacturing industries. The remaining part of the chapter is as follows.

Section Two provides an overview of the IoT and its usage in the manufacturing industry.

Section Three discusses a case study of an industrial sector where IoT is being used.

Section Four presents the current challenges in using IoT for modern-day manufacturing industries and future directions.

Section Five presents the conclusion of the discussion.

4.2   IOT IN MANUFACTURING INDUSTRIES: AN OVERVIEW

IoT is simply a closed-loop control system in which the computing machines, digital machines, and objects are being provided by unique identifiers (UIDs), which has unique identification according to the data transfer through a serial or parallel mode of communication. It reduces the human effort which was being utilized for arranging the machine data logs. An advanced principle that sees industrial equipment in the form of smart manufacturing objects sensing, connecting, and communicating with each other to carry out manufacturing tasks is known as IoT-enabled manufacturing [5]. Information technology's infrastructure for data acquisition and sharing have increased the efficacy of a manufacturing system, which is known as Industry 4.0, a modern manufacturing concept.

Smart manufacturing incorporates IoT enabled technology. It is imperative for manufacturing businesses to conduct real-time data collection and sharing among various resources such as machines, workers, materials, and jobs [6]. Industrial Revolution 4.0 has enhanced the manufacturing sector, which is now entangled with automation robotics, control, and optimization. Integrated manufacturing is a pressing matter. IoT in this field is now capable of dealing with complex decisions [7–11].

The development of programmable logic controllers (PLCs) and supervisory control and data acquisition (SCADA) systems has increased the ease of communicating and data transfer. Decisions can be made automatically, and actuators can perform accordingly. Robots can use their artificial intelligence to optimize the solutions. Previously the whole plant needed to be shut down if any fault occurred. Many times, workers first had to find the fault, and it could take days. As a result, production suffered a lot. Now the identification of fault and its precise place can be found easily by IoT. Actuators now can quickly rectify the fault, but there is still a vacant space for advancements in rectification.

Optimization of cost and raw materials needs to be done. Usually, evolutionary algorithms such as ant-colony algorithm, genetic algorithm, particle swarm optimization, and artificial neural network are being implemented to predict optimized solutions.

IoT can play a vital role in product quality. The controllers and the linkage from the machine to the work manager through IoT like cyber networking or SCADA system keeps him updated. The quality affects the production rate. The faulty workpiece can create a massive disturbance in terms of cost and time in the manufacturing and supply chain becomes a hurdle in restoring the industry's reputation. If a workpiece is going through automatic checking, it will save time. IoT is responsible for retaining the reputation by installing controllers and continuous data transfer from a machine to an end-user. The quality check involves the overall layout and the process information through which an item pass.

The logbook or the statistical workbook has the features of storing the data, which is old fashioned now with the advent of the IoT. Rapid manufacturing needs to get noticed by the manufacturing stations about the quantity and quality of the subject being delivered to the society. The manager is well informed about this IoT way of storing the data. The registers in microcontrollers or microprocessors are being used to store the information, and this data is being saved on the cloud if the PCs must be shut down when the user is not there with it. So the user is well aware of any situation or process being carried out during the batch or continuous type of production, and he/she may act accordingly. Therefore, the informed decisions capability is yet another feature in dealing with an industrial process.

The quality of manufacturing industries has increased significantly after the addition of IoT. With this, it is possible to improve the quality of products, resulting in decreased waste, lower costs, increased customer satisfaction, and increased sales [12].

4.3   IOT BASED SMART MANUFACTURING

Usage of radio frequency identification (RFID), wireless sensor network (WSN), and cloud computing is growing in the modern-day manufacturing industries [10,11,13–17]. A production management system that incorporates RFID was put in place to increase manufacturing flexibility for the assembly of motorcycles [18]. Loncin Motor Co., Ltd. uses this manufacturing system to collect real-time production data from raw materials to improve the visibility, traceability, and trackability of items of interest. Yang et al. describe how Guangdong Chigo Air Conditioning Co., Ltd. implemented RFID-based real-time shop-floor material management [14]. In this case, RFID enabled real-time object visibility and traceability. The use of multi-agent technologies enables the use of parallel robots equipped with distributed agents that have been implemented using an agent-based architecture, thus facilitating intelligent manufacturing [1].

In South Korea and Sweden, manufacturing firms are currently testing and developing ideas for the *smart factory*. One of the overall goals of the investment in the manufacturing sector is to create a world-class manufacturing sector. One good example of a smart factory is a company that makes engine pistons for vehicles and aspires to manufacturing excellence. Molding, casting, machining, and assembly are the major processes. A static mold casting process with accurate dimensional control is required for mass production since engine pistons have inherent properties that make a permanent mold casting process impractical [19]. A digital twin model, which can evaluate and predict productivity, logistics, and quality errors, is developed and implemented using smart sensors, IoT, and big data analytics.

BC Machining Company is using smart approach in determining the life of their machine tools. The constant monitoring of material removal associated with tool determines the life of the tool. IoT acknowledges the perspective of prediction in this case. The data is being stored for the number of operations and its life can be estimated. This company has addressed the matter of tool wear and breakage can decrease the production rate. They have introduced the software namely machine metrics predictive. This diagnosis predicts and hence prevent the machines and raw materials from whole breakdown in production process.

The factory limited like Alps Electric has the tendency to build small electronics with robust designs, meeting today's demand of energy conservation. The interlinking of its franchise is best representation of monitoring and controlling the actuators through fault detection systems. Its sensor and automation concept is flawless, and thus the aim justifies itself as 'perfection of electronics.'

The RFID advancement in defense sector has enhanced IoT applications and Brazilian company is famous for its up-to-date technology. National Center for Advanced Electronics Technology (CEITEC) is working hard in reorganization of human, medications, animals, and blood products. This company is being considered as important partner in strategic planning. It includes the highest ever IT safe passwords for its security and data protection. It has won many certificates in terms of cyber security.

The other industries where the IoT have made a tremendous contribution are health care management [20,21], transportation and logistics [22], firefighting [23,24], and mining [25], to name a few.

4.4 CHALLENGES FOR MANUFACTURING INDUSTRY

IoT is continuously working for the quality and process of the product, but implementation is still gradual [13]. There are three significant challenges identified [14]: one is the general hesitancy of machine builders and end-users to contrivance this technology, second is the security issues, where the collected information about the industry has to be made as big data and to be shared with the co-partners for making it IoT cultured organization, lastly, partnering the other industries, where only a few industries make it possible to bring all the supporting firms into one coherent package [15].

In manufacturing, each vertical sector (aerospace) has different processes and sensors requiring various gateways and platforms to collect the data. There are numerous challenges connecting all the devices, necessitating altering the communication protocol and various underlying technologies [12].

Machines have segments, and each segment is being functioned through a vast number of sensors. Communication on an advanced level is required between these whole structures. Fast internets are there. Presently the centralized paradigm is used to collect the data from different nodes in the network. It needs many secure data servers operational, and the sensitivity level needs to be checked and verified. Therefore, control system-based devices are an option. Working on such devices requires strong expertise and it still needs an operator to observe and alter the functions of actuators where a change is inevitable.

The IoT is a very complicated heterogeneous network because of the connection between various networks using various communication technologies. In the current situation, there is not yet a widely accepted standard common platform that is designed to hide network or communication technology heterogeneity while providing a naming service for a wide range of applications. In managing connected things, the problem is how to promote collaboration between entities and administering devices on a global scale while addressing, identifying, and optimizing devices at the architectural and network levels.

The issue faced by increasing IoT systems is security as the information is continuously shifting to the clouds. IoT has loopholes while addressing securities and privacies as the data transfer rate between the industries and the cloud is enormous. It has put a question mark for their securities. To avoid hacking, continuous changing and strategy algorithms can enhance the capabilities of IoT when dealing with security issues. Data retrieval is the process in which a user can easily get his data whenever required. This is only possible if the data is being transferred as a copy to any cloud-based system, and in that way, robust control can be executed for stakeholders by the managers. While dealing with this, data security needs to be addressed. Smart factories develop intensive data security checks on their websites, clouds, or other data storage systems. Different software and frameworks are being developed by the researchers for ensuring privacy in the storage systems.

HMI has an important feature of connecting the operator with the machine processes. In the task of learning, giving allocations and HMI formations, applications based on mobile are suitable for the management of industrial control systems. The specialized architecture, namely no effort rapid development (NERD), is formed to establish high functionality-based HMI. The advantage of IoT being used here is rapid communication between the operator and the real-time manufacturing process. From time to time, information about the process leads toward smart manufacturing, which eventually has the benefits of maintenance, quality assurance, and safety management of a successful organization. Future work in the IoT based manufacturing industries is being carried in the following domains:

As the level of automation is increasing substantially in the manufacturing industries, it is desired that the efficiency of the process should also increase. The decisions in the automation are carried out by using intelligent algorithms. New advanced algorithms are being studied and developed to minimize disruption and enhance overall efficiency.

Energy consumption and management is another issue that is being focused on for smart manufacturing. A significant portion of the product price covers the cost of energy consumption by the manufacturing industry. With the IoT based industries, the energy management efficiency methods have improved significantly. Various IoT based control strategies are being currently deployed by using various sensors.

One of the significant roles that IoT in manufacturing industries can play is to help in the early detection of equipment maintenance. Various small experimental setups based on sensors and wireless connections can be implemented to get the machine's performance and maintenance data.

4.5   CONCLUSION

The conclusion holds a powerful statement that the IoT emerged in manufacturing as an ingredient for fast production. This will save time, and managers will undoubtedly be at ease. Installation can somehow be complex at the start, but this will pay back all the efforts, and this will settle all the regenerative issues once IoT has been fully installed in every single station in the manufacturing industry. Robotics and automation have certainly reduced human effort. The things being manufactured on a precision scale have their reliability and sustainability long-term claims, and the customer will surely trust those products. The product will have a better future, and suggestions by the end-user are fully implemented for the future.

REFERENCES

[1] Zhong, R.Y., Xu, X., Klotz, E. and Newman, S.T., 2017. Intelligent manufacturing in the context of Industry 4.0: A review. *Engineering*, *3*(5), pp. 616–630.

[2] Wan, J., Chen, B., Imran, M., Tao, F., Li, D., Liu, C. and Ahmad, S., 2018. Toward dynamic resources management for IoT-based manufacturing. *IEEE Communications Magazine*, *56*(2), pp. 52–59.

[3] Wan, J., Tang, S., Li, D., Wang, S., Liu, C., Abbas, H. and Vasilakos, A.V., 2017. A manufacturing big data solution for active preventive maintenance. *IEEE Transactions on Industrial Informatics*, *13*(4), pp. 2039–2047.

[4] Li, Q., Wang, Z.Y., Li, W.H., Li, J., Wang, C. and Du, R.Y., 2013. Applications integration in a hybrid cloud computing environment: Modelling and platform. *Enterprise Information Systems*, *7*(3), pp. 237–271.

[5] Zhong, R.Y., Dai, Q.Y., Qu, T., Hu, G.J. and Huang, G.Q., 2013. RFID-enabled real-time manufacturing execution system for mass-customization production. *Robotics and Computer-Integrated Manufacturing*, *29*(2), pp. 283–292.

[6] Bi, Z., Da Xu, L. and Wang, C., 2014. Internet of things for enterprise systems of modern manufacturing. *IEEE Transactions on Industrial Informatics*, *10*(2), pp. 1537–1546.

[7] Lee, J., Bagheri, B. and Kao, H.A., 2015. A cyber-physical systems architecture for Industry 4.0-based manufacturing systems. *Manufacturing Letters*, *3*, pp. 18–23.

[8] Tan, Y.S., Ng, Y.T. and Low, J.S.C., 2017. Internet-of-things enabled real-time monitoring of energy efficiency on manufacturing shop floors. *Procedia CIRP*, *61*, pp. 376–381.

[9] Li, W. and Kara, S., 2017. Methodology for monitoring manufacturing environment by using wireless sensor networks (WSN) and the internet of things (IoT). *Procedia CIRP*, *61*, pp. 323–328.

[10] Shahbazi, Z. and Byun, Y.C., 2021. Integration of blockchain, IoT and machine learning for multistage quality control and enhancing security in smart manufacturing. *Sensors*, *21*(4), p. 1467.

[11] Wang, Y., 2021. Industrial structure technology upgrade based on 5G network service and IoT intelligent manufacturing. *Microprocessors and Microsystems*, *81*, p. 103696.

[12] Santhosh, N., Srinivsan, M. and Ragupathy, K., 2020, February. Internet of things (IoT) in smart manufacturing. In *IOP Conference Series: Materials Science and Engineering* (Vol. 764, No. 1, p. 012025). IOP Publishing.

[13] Wu, D., Rosen, D.W., Wang, L. and Schaefer, D., 2015. Cloud-based design and manufacturing: A new paradigm in digital manufacturing and design innovation. *Computer-Aided Design*, *59*, pp. 1–14.

[14] Qu, T., Yang, H.D., Huang, G.Q., Zhang, Y.F., Luo, H. and Qin, W., 2012. A case of implementing RFID-based real-time shop-floor material management for household electrical appliance manufacturers. *Journal of Intelligent Manufacturing*, *23*(6), pp. 2343–2356.

[15] Lee, I., 2020. Internet of things (IoT) cybersecurity: Literature review and IoT cyber risk management. *Future Internet*, *12*(9), p. 157.

[16] LD, X., He, W. and Li, S., 2014. Internet of things in industries: A survey [J]. *IEEE Transactions on Industrial Informatics*, *10*(4), pp. 2233–2243.

[17] Abuhasel, K.A. and Khan, M.A., 2020. A secure industrial Internet of things (IIoT) framework for resource management in smart manufacturing. *IEEE Access*, *8*, pp. 117354–117364.

[18] Liu, W.N., Zheng, L.J., Sun, D.H., Liao, X.Y., Zhao, M., Su, J.M. and Liu, Y.X., 2012. RFID-enabled real-time production management system for Loncin motorcycle assembly line. *International Journal of Computer Integrated Manufacturing*, *25*(1), pp. 86–99.

[19] Wiktorsson, M., Do Noh, S., Bellgran, M. and Hanson, L., 2018. Smart factories: South Korean and Swedish examples on manufacturing settings. *Procedia manufacturing*, *25*, pp. 471–478.

[20] Domingo, M.C., 2012. An overview of the Internet of things for people with disabilities. *Journal of Network and Computer Applications*, *35*(2), pp. 584–596.

[21] Pang, Z., Chen, Q., Tian, J., Zheng, L. and Dubrova, E., 2013, January. Ecosystem analysis in the design of open platform-based in-home healthcare terminals towards the internet-of-things. In *2013 15th International Conference on Advanced Communications Technology (ICACT)* (pp. 529–534). IEEE.

[22] Qin, E., Long, Y., Zhang, C. and Huang, L., 2013, July. Cloud computing and the internet of things: Technology innovation in automobile service. In *International Conference on Human Interface and the Management of Information* (pp. 173–180). Springer.

[23] Ji, Z. and Anwen, Q., 2010, November. The application of internet of things (IOT) in emergency management system in China. In *2010 IEEE International Conference on Technologies for Homeland Security (HST)* (pp. 139–142). IEEE.

[24] Zhang, Y.C. and Yu, J., 2013. A study on the fire IOT development strategy. *Procedia Engineering*, *52*, pp. 314–319.

[25] Qiuping, W., Shunbing, Z. and Chunquan, D., 2011. Study on key technologies of Internet of things perceiving mine. *Procedia Engineering*, *26*, pp. 2326–2333.

5 Internet of Things-Enabled 3D Printer

G. Hussain, Abid Imran, Salman Amin,
Wasim A. Khan, K. Rehman, G. Abbas,
Ali Alvi, Ahmed Murtaza, Daniyal Akram
and Roshan Rehman

CONTENTS

5.1 INTRODUCTION

Since the advent of the steam engine, and the following dawn of Industrial Revolution, humanity has constantly been at the forefront of upgrading and advancing technology with the underlying pursuit of achieving two things—efficiency, thus human ease and scalability.

Essentially, the world has seen three industrial revolutions, starting circa 1765 [1]. The first of the three saw the advent of mechanization, with a steam engine, to shift laborious tasks to machinery. To refine the experience, i.e., introduce efficient new sources of energy mainly electricity, gas and oil were introduced but required replacement of the steam engine, which by virtue of its name, ran on steam. The internal combustion engine was born marking the pinnacle of the second revolution in 1870. The third revolution brought forth electronics and computers, around 1969, that bridged communication gaps and allowed usage of machines through digital interfaces and programming.

Quite surprisingly, what we see in modern times is an enhanced version of what the third revolution brought forth. While the hallmark of 1969 and beyond was digitized communication with machines, current efforts are directed to "inter-cell" or "inter-machine" communication through a common channel or portal. This refers to ongoing organic change as Industry 4.0—characterized by the usage of internet and the World Wide Web as the common portal referred to earlier. Machines, in an idealized world governed by the internet of things—colloquially IoT—would allow a triad of communication: human to machine and machine to machine without physical human intervention.

DOI: 10.1201/9781003220985-5

This research focuses on Industry 4.0 with emphasis on establishing a human to machine communication channel through the World Wide Web. The machine cell under work is a 3D printer. The aim is to set building blocks for advent of full-scale implementation of Industry 4.0 through this communication channel thus achieving, as targeted earlier, efficiency and human ease.

5.2 METHODOLOGY

The conclusion reached by reviewing various literary resources prompted us to incorporate an IoT based solution right after the CAD model is generated and exported as an STL file on the end user's machine. Our hybrid strategy will enable us to relay the process of "slicing" the STL file over the internet to produce the G codes needed to drive the relevant printer components. In addition to remote printing, the capability of storing the part file along with its information will be dealt with in the last phase of design where the Amazon Simple Storage Service (S3) [2] will be utilized. The underlying configuration of the printer electronic system allows to connect a micro controller which can be used to expand the functionality of the printer as indicated in Figure 5.1.

Unlike the conventional approach, the requirement of having a physical computer will be disregarded and the wireless lan (WLAN) capabilities of the micro controller will be used to receiving the printing data from the hosting application. The micro controller interfacing will allow the transfer of G codes and establishes parallel communication with the printer motherboard; this will not only allow the control over the printer (i.e., the extruder, the heating bed and motors) but also enables the extraction of sensory and spatial information.

As discussed in the preceding section, it was decided to use OctoPrint as the hosting software responsible for monitoring and control activities, meanwhile, the Raspberry Pi was chosen as the compatible micro controller for our setup. OctoPrint client acts as a webserver, providing a graphical user interface to communicate and control the 3D printer from a web browser. It monitors the printing status and temperatures, furthermore, the use of components that attach with the micro controller such as a camera can provide live feedback of the printing status within OctoPrint interface [3]. This communication between the OctoPrint webserver, the camera module, the 3D printer and the relays to turn the printer on/off is depicted in the Figure 5.2

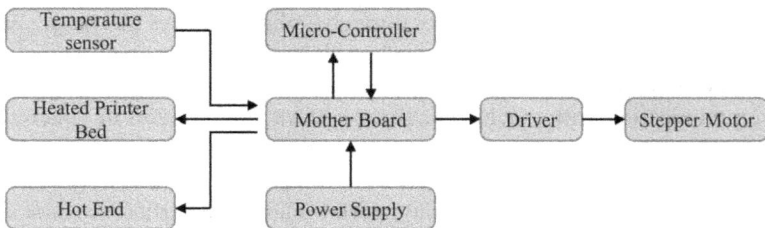

FIGURE 5.1 3D Printer Electronic Diagram, Configuration Showing Controller Communication with the Printer's Electronic Components.

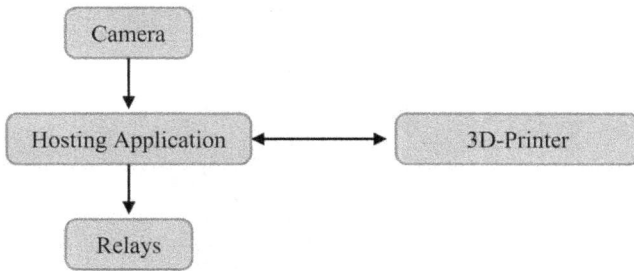

FIGURE 5.2 Remote Webserver Communication Diagram, Communication between a Hosting Platform (OctoPrint), Camera Module, 3D Printer and Relays to Turn the Printer On or Off.

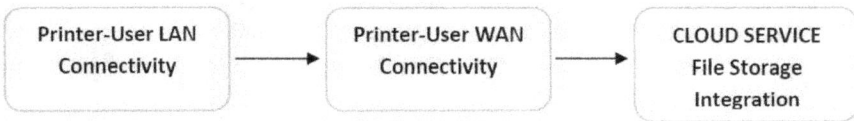

FIGURE 5.3 Design Process Workflow, Shows the Workflow from Local Area Network to Wide Area Network and Ultimately to the Cloud Service for File Storage Integration.

below. It is worth noting that other services, such as Vision Authentise [4], required a subscription on a website and alternative hosting software such as Astroprint [5] did not possess enough developer plugins and troubleshooting documentation.

A drawback of using OctoPrint out of the box is its nature to work on local area network and an inability to perform the slicing operation on the model's STL file.

Since the objective of this research was to minimize human intervention and to perform the printing remotely over the World Wide Web, it was decided to execute the project in three phases. First, set up OctoPrint on the local area network. The second phase will involve migration from the local area network to the World Wide Web and the third phase will focus on integrating cloud service into workflow for the scalability of the project. (Figure 5.3).

5.3 NETWORK ARCHITECTURE

It starts with burning an image of the OctoPi [6] firmware onto SD card which will create an instance that contains and runs OctoPrint, and a live video streaming plugin called mjpg streamer [6]. When coupled with the Raspberry Pi connected with the USB port of the printer it will allow parallel information flow. This preliminary setup will conclude the first phase of objectives and will allow the transfer of data over the webserver created by the OctoPrint over the local area network by the means of a static IP that a user will input into their desktop or mobile web browser that relates to the same network as the printer. This process of inter communication on the local network is briefly illustrated in (Figure 5.4) [7].

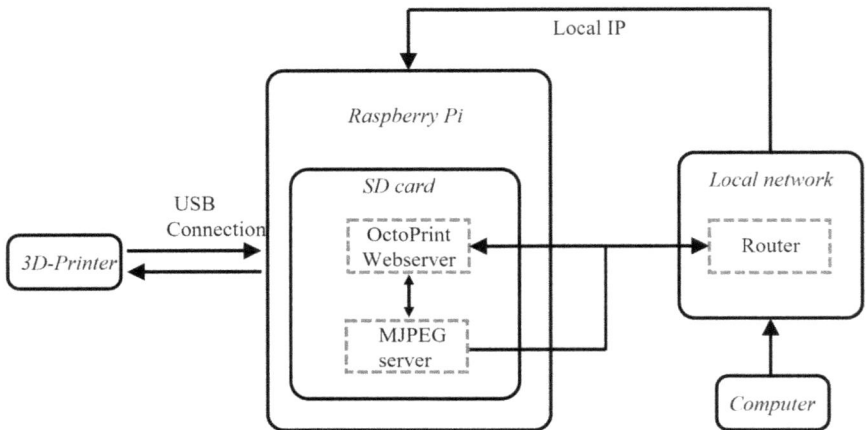

FIGURE 5.4 Local Network Architecture Diagram Shows Inter Communication of the 3D Printer with the OctoPrint Webserver and Local Area Network Using Raspberry Pi Controller.

A local area network (LAN) consists of access points that allow devices to connect to web servers and internal servers within a single building, campus or private network [8]. Since the devices are on the same network, they are assigned an IP address by the router and so they can share files and data over the shared network. Using the system of LAN is quite straight forward and serves as a secure option to print by simply sending the unsliced SLT file over to the OctoPrint via the web interface. Alternatively, the SLT file can be sliced on the user's computer by a software like Cura [9] and then sent over for printing. Each device connected to the router has its own ports if the Raspberry Pi has an assigned IP of 192.168.1.50 and a communication port ninety.

A user's computer will have to ask the router to grant access to the device identified by

192.168.1.50:90 to consequently gain access to the webserver of OctoPrint using their internet browser. This marks the completion of the first phase of the project. However, it does not meet the demand of true IoT integration as the team targeted remote web access. This approach is viable only if someone would like to sit in their respective offices and decides to print a part.

The second phase of this study focuses on granting access to someone off campus or even abroad. For this there need to open the communication ports of the devices to the public; thus, port forwarding must be configured. The Raspberry Pi will have to be assigned a public port, i.e., two hundred, thus the router will know that the device with the local identifier 192.168.1.50:90 is 'tagged' with the public port number of two hundred so any external requests made to the router by its public IP ending with the port number two hundred, e.g., 14.34.507.983:200, will ask it to communicate with the Raspberry Pi connected to the router through the local network. Figure 5.5 shows this relationship process.

The Figures 5.4 and 5.5 represent the processes where a microcontroller intervention acts to extend the functionality of the printer from here an iterative process will be adopted to achieve the objectives of the design project.

It can be observed from the Figure 5.5 that a local device such as the Raspberry Pi cannot communicate to an external computer using its local IP. Instead the public IP, also called the WAN IP address, of the router is used. This information can be obtained using any website such as WAN IP [10] when connected to the local area network. It is worth noting that making a device accessible over the internet carries risk, however, this can be tackled by proper encryption mechanisms or by using a VPN a software that masks IP addresses. Fortunately, the OctoPi image contains an encryption software that acts as a proxy, called haproxy, which will require authentication for every request made to OctoPrint otherwise it will not grant access. This configuration with the public IP in practice is illustrated in Figure 5.6.

With this port forwarding in place, printer can be accessed remotely through any network, anywhere in the world just as one would locally. This configuration concludes the second phase of the design.

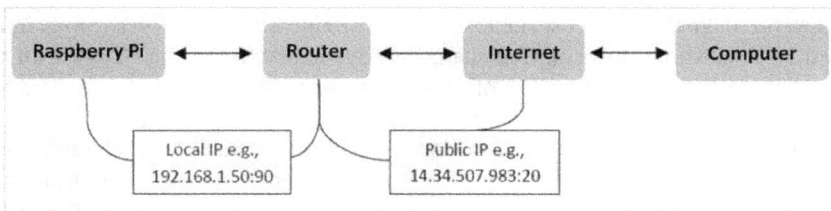

FIGURE 5.5   Local IP Correspondence with Public IP Shows the Communication of the Controller with Router on Local Area Network and a Computer over the Internet.

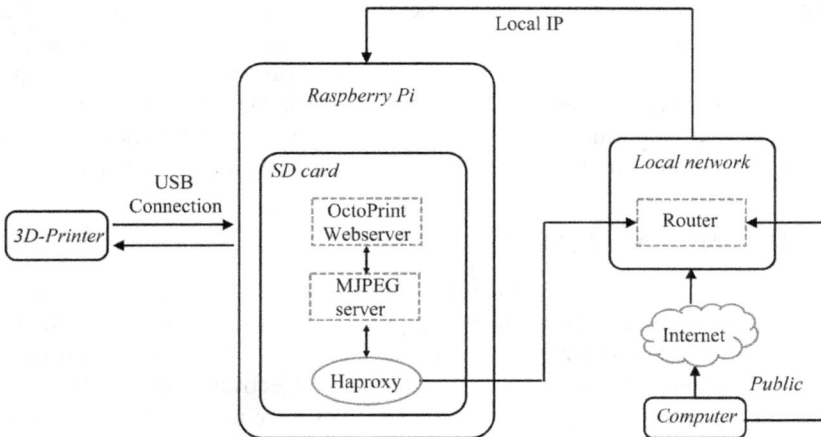

FIGURE 5.6   Remote Access Network Configuration Shows the Configuration of the 3D printer with the Controller and a Remote Computer through Internet.

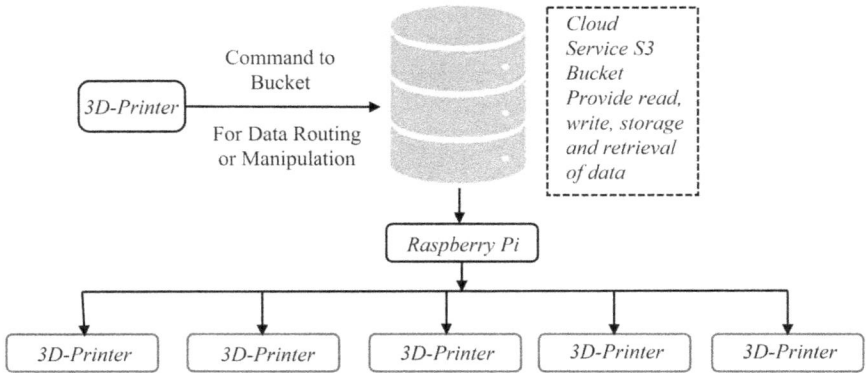

FIGURE 5.7 Cloud service simple storage integration shows the route of information flowing from computer to a 3S cloud service and then to the controller which is providing information to different 3D printers.

Now, to make the project scalable, Amazon Simple Storage Service (S3) is introduced as a substitute to using the scarce storage available on the SD card of the micro controller. The integration of cloud computing S3 can be of immense use when deployed on a manufacturing cell composed of

3D printers that work in correspondence with other cells all using the same cloud storage. The user's part file and information will be stored onto the cloud from their operating machine and fetched remotely over by a single or a group of 3D printers according to requirement. This factor of cloud interconnectivity is a vital step in manufacturing automation and is widely used in sophisticated manufacturing and assembly plants such as Tesla's giga factory.

The network architecture in this phase will be the same as that observed in Figure 5.5 where the only difference lies in the how the "internet" is utilized. Figure 5.7 describes the route of information flow if cloud services S3 is in place. Previously, it was used for routing data; now it'll be storing the data too which can be later fetched any time when the user needs it. The OctoPrint instance can also be launched multiple times in the backend which can potentially allow controlling multiple 3D printers simultaneously for printing either the same or distinct parts.

5.4 TESTING AND RESULTS

During the initial stages of the research, a range of deliverables were set. These were eventually used as the KPIs for the whole project. The objectives were divided into two major segments—in terms of connectivity, as well as the range features accessible over the internet. To start off, the following objectives were initially set: By the end of the research project, all the deliverables were met. A phase-wise implementation of the project was conducted by first achieving a remote connection to the printer by using a common Wi-Fi router. The features accessible over this type of connection met all the deliverables as mentioned in Table 5.1.

TABLE 5.1
Deliverables.

Connectivity	Accessible features
Secure wireless connection over local Area Network (Wi-Fi)	Tool temperature control
Secure wireless connection over the World Wide Web	Bed temperature control
	Slicing and g-code Generation
	Tool movement along all axis
	Intervening print job
	Progress monitoring

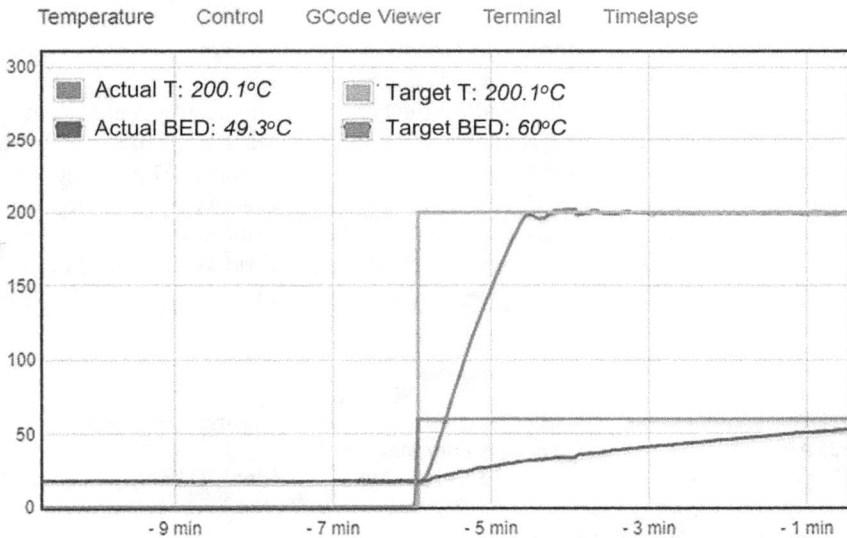

FIGURE 5.8 Temperature Control Shows a Graph between the Step Input of Bed and Nozzle Temperature of Different 3D printers [11].

For pertinence [11], Figure 5.8 clearly shows how a step input of bed and nozzle temperature over the internet is achieved in real time. It takes approximately two to three minutes for the temperatures to achieve the desired value. No problems related to latency were identified in this regard.

Multiple tests were conducted to make sure that desired tool and bed temperatures were achieved. Moreover, the tool was also moved wirelessly along all its axis. It is worth noting that through various tests, it was realized that the latency of giving a command to the tool, and the tool moving was no more than 0.5 seconds over LAN. The platform used in this research gave the user full autonomy to directly upload a 3D model into the platform, and the slicing and g-code generation was done automatically. Moreover, the platform also provided live feed of the print job by attaching a web cam on the 3D printer. Therefore, it was also possible for the user to visually monitor the progress of a print job over the internet. In case anything went wrong, an option to pause or stop the print job was also available.

   After achieving the deliverables on local area network, this process was moved over to the internet by using an open-source plugin. This allowed for all the features mentioned earlier to be accessible over the internet. One of the differences which was noticed when testing the movement of tool over the internet was that the latency increased to around one second. However, this change was negligible, thus, the performance of the overall system seemed acceptable.

5.5   CONCLUSION

The design and development of IoT based 3D printer according to the requirements is presented in this research. The major offering of this study is to establish inter machine communication for optimized industrial manufacturing and quality control through realization of a manufacturing system remotely using the concept of internet of things. "Internet of things" is used for the purpose of connecting and exchanging data with other devices and systems over the internet. This work incorporates the concept of internet of things on industrial scale by designing a 3D printing workstation that can be accessed anywhere via internet. This research covers the implementation of the concept on local area network only, however more work can be carried out on establishing communication channels through World Wide Web. This work can be further progressed to be implemented in the industry.

REFERENCES

  [1] The Arena Group. Available: www.history.com/topics/industrial-revolution
  [2] Amazon S3. Available: https://aws.amazon.com/s3/
  [3] G. MP, Y. Shinde, R. Madaki, and S. Nadaf. (2019). "IoT Based 3D Printer," *International Research Journal of Engineering and Technology (IRJET)*, Vol. 06, Issue 04, www.irjet.net.
  [4] Authentise Data- Driven Workflow Management. Available: www.authentise.com/
  [5] Astroprint. Available: www.astroprint.com/
  [6] OctoPi. Available: https://github.com/guysoft/OctoPi
  [7] Mjpg-streamer. (2021) *GitHub*. Available: https://github.com/jacksonliam/mjpg-streamer
  [8] Heavy AI. (2021). *Local Area Network*. Available: www.heavy.ai/technical-glossary/local-area-network
  [9] Ultimaker Cura. Available: https://ultimaker.com/software/ultimaker-cura
 [10] Wanip.info. Available: http://wanip.info/
 [11] MathijsG. (2018). Available: https://github.com/OctoPrint/OctoPrint/issues/2437

6 Collaborative Robot with Collision Avoidance System

*Ijlal Ullah Khan, Muhammad Umair Shafiq,
Abid Imran, Arsalan Arif and Wasim Ahmed Khan*

CONTENTS

Human involvement has almost been replaced by the machines after the Fourth Industrial Revolution. In Industry 4.0, the IoT based communication is being used between different nodes, which includes different machines and mobile robots, etc. The optimal solution with precision and accuracy can effectively be achieved with the involvement of fully collaborative robot (COBOT) and artificial intelligence in Industry 4.0. Considering the requirement of Industry 4.0, this chapter deals with the design, implementation, kinematic modeling of COBOT and optimal path planning algorithms for object handling between different nodes.

DOI: 10.1201/9781003220985-6

6.1   BACKGROUND AND MOTIVATION

Pakistan is a growing economy and derives 20% of its gross domestic product (GDP) from the industrial sector, with just 7.6% coming from the secondary sector. The primary sector of Industries, which includes raw materials and unfinished items, contributes 12.4%. The reason for this is because Pakistan continues to rely on human labor rather than bringing contemporary technologies into the industrial sector. It would not only result in lower finished products costs as well as production times but also in higher quality goods/items. The demand for innovation in Pakistani industries provides us the impetus to develop a COBOT; which transports any tool or equipment to a desired location with the ability to avoid any unknown obstacles along the way.

6.2   LITERATURE REVIEW

COBOT is a type of robot capable of moving between multiple nodes to transport various goods to avoid the obstacles as well as following the shortest path. It is really a piece of mechanical gear capable of transporting components between machines and cells. COBOTs are also known as collaborative robots since they interact with humans in a shared area. COBOTs are self-aware, articulated robots.

Many industrial collaborative robots have been developed so far which are working for different application in different environments. Amazon Robotics produces mobile robots [1], which reads electrical bar codes on the floor to navigate while avoiding any collision with obstacles. Amazon warehouses now employ over 200,000 mobile robots with hundreds of thousands of humans [2]. All the robot and human are working together in a dynamic environment. Kiva has a robot named Kiva Amazon DU 1000, which is 75 cm long, 60 cm wide and 30 cm high. Its weight is 110 kg and can travel with five km/h. Otto Motors builds industrial collaborative robot equipped with safe navigation in complex situations [3–4]. Frederic Ladinos [5] ER is a COBOT platform from Easy Robotics. ER COBOT platform is also equipped with a robotic arm. Kiwibot [6] is a four-wheeled robotic platform, which can carry four to five orders at a time and navigates customer within 300 meters proximity with an average delivery time of 27 minutes. Kiwibot [7–10] has successfully delivered many mobile robots which uses artificial intelligence (AI) and machine learning and can move in sideways by using mecanum wheels. Kiwibot are not fully autonomous, they are semi-autonomous bots. The delivery bots are equipped with a camera to detect obstacles while navigating in a dynamic environment [11].

To date, different research activities have been carried out for the path planning of mobile robot in dynamic environment considering the static and dynamic obstacles. The Dijkstra algorithm is proposed by E. W. Dijkstra in 1959 [12] which is a typical shortest path algorithm for solving the shortest path problem in a directed graph. The main feature of this algorithm is that it can consider the weight of all the nodes to calculate the distance from start point to goal point. Hart et al. [13] proposed A* algorithm in 1968. The A* algorithm is developed based on the Dijkstra algorithm. Starting from a specific node, the weighted value of the current child nodes is updated, and the child node which has the smallest weighted value is used to update the current node until all nodes are traversed. Furthermore, kinematic modelling of mobile robot is done to follow the desired trajectory of mobile robot. There are various types of mobile robot which can be used to follow the trajectory and each mobile robot has its own kinematic model based on the number of wheel and type of wheel. Eric N. Morit et al. [14] have developed the kinematic model of a car like mobile robot. The author has described the velocity vector and calculated the relation of velocity vectors with each wheel. Wei Ren et al. [15] have developed the mathematical model of a car like robot to follow the desired trajectory.

This chapter presets the complete design of COBOT following by the kinematic modeling of the COBOT to follow the desired trajectory. Moreover, the power, force and torque analysis are performed to calculate the required torque for each motor. The proposed design required four motors while two middle wheels are passive. Furthermore, stress analysis is performed in ANSYS, and complete computer aided design (CAD) model is designed using solid works. The dynamic simulations are performed in MSc Adams to analyze the kinematic and dynamic of the designed COBOT. Finally, the A* path planning algorithm is used to find the desired static collision free trajectory of the mobile robot. This trajectory is then used to move the COBOT from initial position to goal point.

6.3 KINEMATIC MODELING AND MOTOR TORQUE CALCULATION

In this section, firstly, the kinematic modeling is performed. Moreover, the power, force and torque calculation are performed to find the required torque for each motor by considering the pay load along with the weight of the robot as 35 kg.

6.3.1 KINEMATIC MODELING AND ANALYSIS

Mobile COBOT is designed to move in any direction. The proposed COBOT consists of total six wheels, in which four wheels are mecanum and two are omni-directional wheels. Mecanum wheels are actuated by the DC motors while omnidirectional are free to move. Kinematic modelling of the mecanum wheels is done to calculate the angular velocity of each mecanum wheel. Figure 6.1 shows the kinematic details of the mecanum wheel along with its velocity vectors [16].

The Kinematic model of the mobile robot is derived from the movement vector of four-wheel drive mobile robot platform. Robot velocity vector v which has parallel direction to the x coordinate of v_x and v_y vector components can be written as follows [16].

$$v_x = vcos\theta, \tag{6.1}$$

Similarly, for y-component,

$$v_y = vsin\theta, \tag{6.2}$$

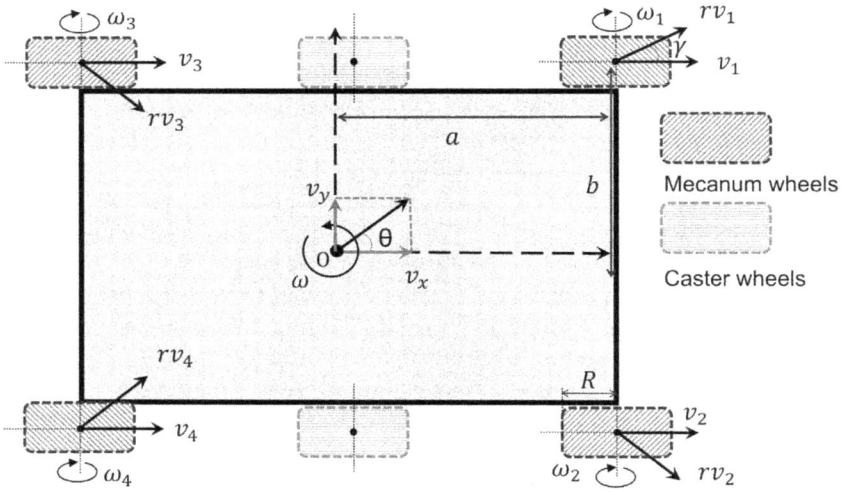

FIGURE 6.1 Movement Vector and Coordinate System of Four-Wheel Drive Mobile Robot Platform [16].

and the angular velocity can be written as

$$\omega = \frac{v}{b},\tag{6.3}$$

where θ is the lateral direction angle of robot movement velocity. Robot measurement is indicated by radius of 'a' and 'b' between body center of robot and wheel hub of $a_i = \{a, a, -a, -a\}$ and $b_i = \{b, -b, b, -b\}$ where $i = \{1, 2, 3, 4\}$ denotes the actuated wheel numbers as shown in Figure 6.1. Linear velocity vector of the wheel and velocity of mecanum roller heading of each wheel is demonstrated by v_i and rv_i, individually. Tilted point γ between v and rv is 45° which speaks to the mecanum roller point each wheel of $\gamma_i = \{\pi/4, \pi/4, -\pi/4, \pi/4\}$. The velocity vector equation of the mobile robot toward coordinate system component can be calculated by

$$v_i + rv_i \left(\cos\gamma \right) = v_x - b_i\omega ,\tag{6.4}$$

similarly, for y-component, the expression will be

$$rv_i \left(\sin\gamma \right) = v_y + a_i\omega \tag{6.5}$$

by rearranging the equations (6.4) and (6.5), we get the following equation

$$v_i = v_x - v_y - a_i\omega - b_i\omega.\tag{6.6}$$

Putting the values of i, a_i, b_i in equation (6.6), the linear velocity equation for each mecanum wheel can be written as follows

$$v_1 = v_x - v_y - (a-b)\omega, \tag{6.7}$$

$$v_2 = v_x + v_y + (a+b)\omega, \tag{6.8}$$

$$v_3 = v_x + v_y - (a-b)\omega, \tag{6.9}$$

$$v_4 = v_x - v_y + (a+b)\omega, \tag{6.10}$$

where the angular wheel velocities are given as $v = \omega R$ and R denotes the radius of the each mecanum wheels. It is assumed that each wheel has the same radius. Equation (6.7–6.10) can be further modified as follows

$$\omega_1 R = v_x - v_y - (a-b)\omega, \tag{6.11}$$

$$\omega_2 R = v_x + v_y + (a+b)\omega, \tag{6.12}$$

$$\omega_3 R = v_x + v_y - (a-b)\omega, \tag{6.13}$$

$$\omega_4 R = v_x - v_y + (a+b)\omega. \tag{6.14}$$

Taking Common of R, v_x, v_y and ω from equation (6.11–6.14) the relationship to find the angular velocity of each wheel for the given vx, vy and ω is established as follows

$$\begin{bmatrix} \omega_1 \\ \omega_2 \\ \omega_3 \\ \omega_4 \end{bmatrix} = \frac{1}{R} \begin{bmatrix} 1 & -1 & -(a+b) \\ 1 & 1 & (a+b) \\ 1 & 1 & -(a+b) \\ 1 & -1 & (a+b) \end{bmatrix} \begin{bmatrix} v_x \\ v_y \\ \omega \end{bmatrix}, \tag{6.15}$$

where ω_1 to ω_4 shows the angular velocities of each wheel. By using the developed model in equation (6.15), one can calculate the required angular velocities of each wheel for the given input of v_x, v_y and ω.

6.3.2 FORCE CALCULATION

In this section the power consumption by each DC motor is calculated. The power is calculated considering the combined payload and weight of the robot as 35 KG.

From Figure 6.2, according to the Newton's law of gravitation

$$F_{Normal} - W_{net}' = 0.$$

Which can be written as follows

$$F_{normal} = W_{net}, \tag{6.16}$$

where, considering the weight of the robot as 35 kg

$$W_{net} = m \times g = 9.81 \times 35 = 345\ N = F_{normal}. \tag{6.17}$$

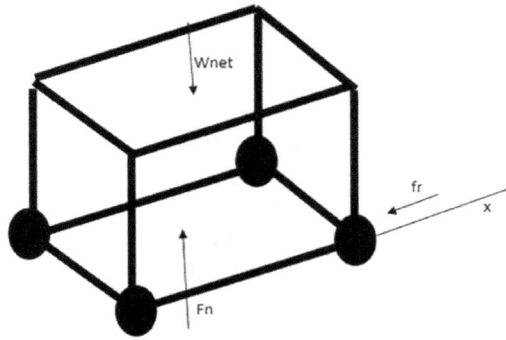

FIGURE 6.2 Free Body Diagram of Forces on COBOT Body.

Similarly for horizontal direction when the robot is initially at rest. All the forces on x-axis will be zero such as $\sum F_x = 0$

$$F_{thrust} - f_r = 0, \tag{6.18}$$

where $F_{thrust} = f_r$ for constant speed and further one can write

$$F_{thrust} = \mu_s F_{normal} = 0.24 \times 345 = 83 N, \tag{6.19}$$

where F_{thrust} is the thrust force when robot will be at rest, μ_s is the static coefficient of friction.

When the robot will be accelerating the required force will be calculated by using the Newton's second law

$$\sum F_x = ma. \tag{6.20}$$

Equation (6.20) can be written as

$$F_{thrust} - f_r = ma, \tag{6.21}$$

where $f_r = \mu_k F_{normal}$. μ_k is the kinematic coefficient of friction.

Finally, the thrust force while accelerating can be calculated as follows

$$F_{thrust} = f_r + ma = 135 \ N \ . \tag{6.22}$$

6.3.3 POWER CONSUMPTION

Based on the force calculation in section 2.2, one can calculate the power consumption as follows

$$P = F_{thrust} \cdot V_{nominalSpeed} + P_{motorlosses}, \tag{6.23}$$

where the total power is calculated as follows

$$P_1 = 40.5\ Watts\left(case-I\right), P_2 = 67.5\ Watts\left(case-II\right). \tag{6.24}$$

Where rest and accelerating states of the COBOT are represented by case I and case II, respectively.

6.3.4 REQUIRED TORQUE CALCULATIONS

In this sub-section, torque required for each motor is calculated for both cases.

$$T_{PerWheel} = F_{PerWheel} \times r_{wheel.} \tag{6.25}$$

where $r_{wheel} = 55$ mm. Thus, the force required by per wheel is give as follows

$$F_{PerWheel} = \frac{F_{Thrust}}{4F_{PerWheel1}} = \frac{83}{4} = 20.8N\ with\ constant\ speed, \tag{6.26}$$

and

$$F_{PerWheel} = \frac{135}{4} = 33.75N\ when\ accelerating. \tag{6.27}$$

Finally, the total torque required is calculated as follows

$$T_{PerWheel_1} = F_{PerWheel} \times R, \tag{6.28}$$

where $T_{PerWheel_1}$ is the torque of each wheel in rest position. $F_{PerWheel}$ is the force acting on each wheel and R is the radius of each wheel.

$$20.8 \times 0.055m = 1.144\ Nm\ with\ constant\ speed. \tag{6.29}$$

Considering the factor of safety as 6, one can write

$$T_{net_1} = T_{PerWheel_1} \times 6 = 6.864 \approx 7Nm\ with\ constant\ speed. \tag{6.30}$$

Equation (6.30) describes the net torque acting on each wheel. Next, one can find the torque for case II as follows

$$T_{PerWheel_2} = F_{PerWheel_2} \times r_{wheel} = 33.75 \times 0.055m = 1.85Nm, \tag{6.31}$$

similarly, considering the factor of safety as 6, one can write

$$T_{net_2} = T_{PerWheel_2} \times 6 = 11.1\ Nm\ when\ accelerating. \tag{6.32}$$

Equation (6.32) describes the net torque of each wheel while accelerating.

TABLE 6.1

Weight Property Index for Different Material Rods.

Material	Thermal expansion Coefficient (10^-6) WF=0.333	β (%)	Density (kg/m^3) wf=0.167	β (%)	Cost (Rs./kg) wf=0.333	β (%)	Strength (MPa) wf=0.167	β (%)	Weight property index (WPI)
Aluminium	23.6	72.03	8700	91.95	256	26.95	700	100	65.02
Stainless steel	17.0	100	8000	100	69	100	414	59.14	93.18
Copper	17.1	99.42	8960	89.29	950	7.26	350	50	58.79

6.3.5 MATERIAL SELECTION

Weight property index method [17] is used to finalize the material of the COBOT. Higher the value of weight property index (WPI) better will be the material selection option. For density, cost, and thermal expansion coefficient:

β (%) = {(lowest value under consideration)/(numerical value of property)} x 100

For strength:

β (%) = {(numerical value of property)/(largest value under consideration)} x 100

Weight property index of stainless is higher than aluminum and copper that's why stainless steel has been selected.

6.4 DESIGN AND SIMULATION ANALYSIS

In this section the CAD model design of the COBOT along with stress analysis and dynamic simulations are presented.

6.4.1 GEOMETRIC MODELLING AND DESIGN

The detailed CAD modelling of COBOT using Solid Works is presented in this sub-section as shown in Figure 6.3. The aim of the COBOT is to store the object by picking from one machine and then place it by reaching another machine. Thus, considering the objective of the COBOT CAD model is designed in such a way to store the required objects. The pay load of the COBOT and revolute robot which will be mounted on this COBOT is calculated and accordingly the COBOT is designed and assembled by using Solid Works. The main parts of the COBOT are following. First, a 3D sketch of the upper tray s designed. After designing the 3D model of the tray, pipes and weldment feature is used to for the 3D model of the cover of tray. Tray is sketched stepwise by selecting respective planes. After that, trim extrude command is used at ends of pipe joining for the proper finishing of

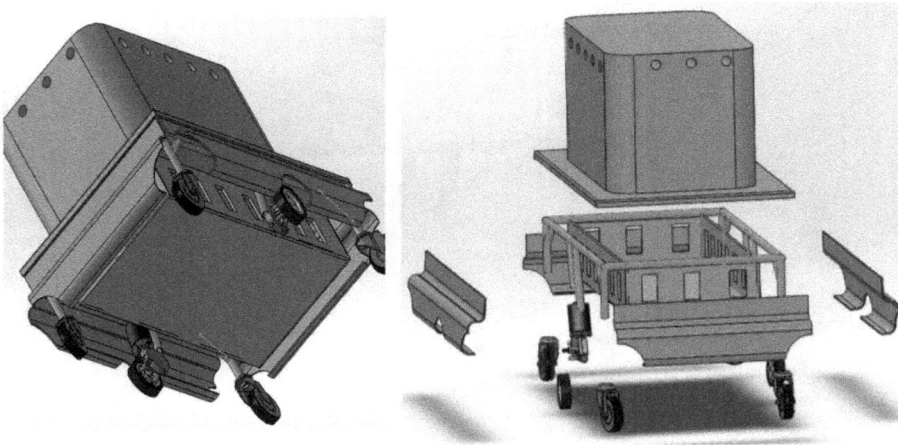

FIGURE 6.3   Final CAD Model of COBOT.

tray cover. And then sketching and cut extrude feature is used to make the slots in the tray on all the side and by dragging the cut extrude on each side to equidistant slots feature on all the sides. After completing the assembly, the next step to design the wheel on the base. Total six wheels are used in this COBOT. Four mecanum wheel which will be actuated by using the DC motors and two omnidirectional wheels are used in the middle which will support the COBOT. The parts are assembled in the sequence that tray is mated to the wheels. After making the lower portion of COBOT, the mating feature of concentric is applied on the points where wiper motors and wheels get attached to the pipes. Distancing in the mating feature helped to fix the assembly of the guards with the tray. After bringing together the lower part of the assembly, simply the sketch in made on the top plane and then flanges are made in the upper part. It is further processed by the extruding of the sketch and then the final touches were applied. This model can now be exported to a CAD modelling software such as Creo Parametric, Solid Works, or others for additional processing

6.4.2 ANALYSIS

In this section, stress analysis is performed to calculate the axial and bending stress. After stress analysis, displacement of the COBOT is also calculated. The axial stresses that generated due to applied load of 343.45 N has been depicted in Figure 6.4 below. The bending arises compressive and tensile stresses according to flexure formula $\sigma_b = My/I$. Where y indicates the distance from the neutral axis and I moment of inertia about the axis of bending, the compressive at the top of the frame and tensile at the bottom. These compressive and tensile stresses get added to axial forces along the beam axis. This overall phenomenon determines the stress propagation in the beam.

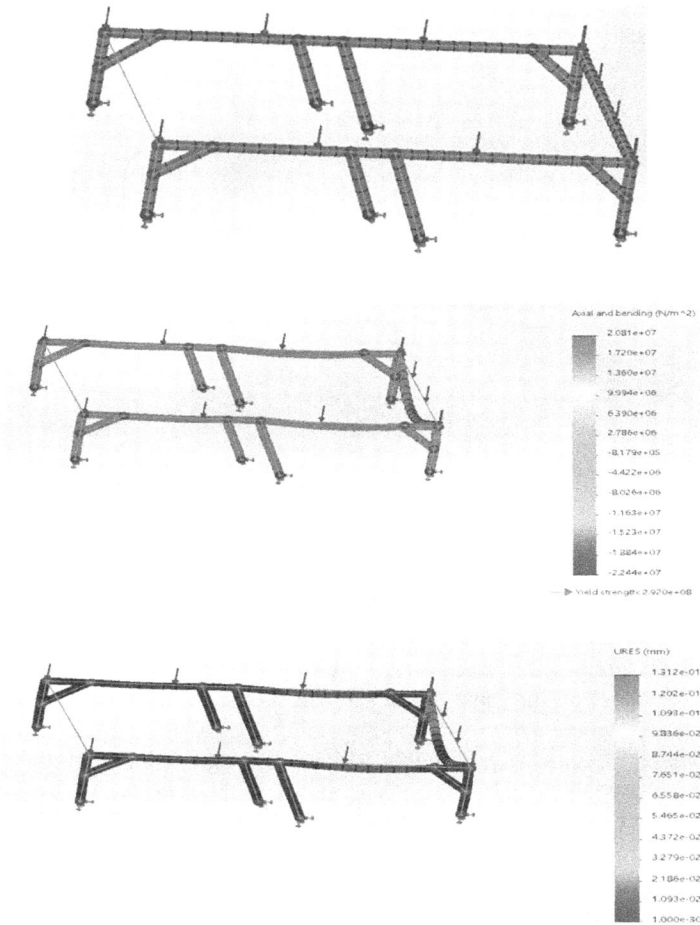

FIGURE 6.4 (a) Free Body Diagram of Load and Reaction Forces on Right Side Beam (b) Shear Force Diagram (c) Bending Moment Diagram.

6.4.3 STRESS ANALYSIS FOR INNER FRAME

The tetrahedron is utilized in 3D finite element analysis, which simplified the total computation. The mesh size of 13.81 mm is used in a moderate range to avoid overburdening

The computer processing unit. The resulting result is an approximation. The axial tensions created by the 343.45 applied load are illustrated in Figure 6.4. According to the flexure.

$$\sigma_b = \frac{My}{I}, \tag{6.33}$$

Where y is the distance from the neutral axis, and I is the moment of inertia about the bending axis. Axial forces along the beam axis increase compressive and tensile strains. Globally, this influences the stress propagation in the beam. Similar pattern is shown in Figure 6.4. We consider it as a beam since the length to thickness ratio is higher. The displacement occurs along the frame when loaded. Figure 6.3b shows the displacement range. Aside from the side frame's maximum displacement, the front frame's overall benign displacement is owing to the overdesign three-wheel support. Plot shows greatest and lowest displacement over the frame. Material and physical qualities of design idea have been kept modest to prevent unnecessary expenses and overdesign. Stainless steel is a ductile material. The maximum von Mises stress criterion [18] was used to the design frame model to forecast ductile material failure.

6.4.4 MSc Adams Simulation

Based on the CAD model, which is developed using the Solid Works, numerical simulations for dynamics analysis is performed in MSc Adams to analyze the designed COBOT. Various kinematics and dynamics analysis is done in MSc Adams to follow the trajectory. To analyze the study of dynamics and moving parts, dynamic simulations are carried out in this section. By actuating all the motors according to the given desired trajectory, the motion of the COBOT is analyzed. The dynamic simulation results are shown in Figure 6.5.

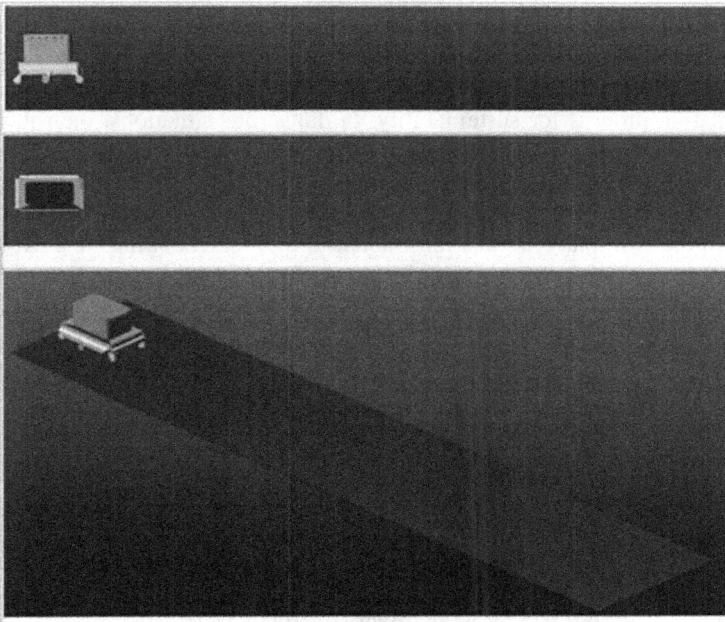

FIGURE 6.5 Dynamic Simulations in MSc Adams.

6.5   IMPACT AND ECONOMIC ANALYSIS

6.5.1   SOCIAL IMPACT

Social impact refers not only to the degree to which the local community is affected but also considers the working class and the key stakeholders. COBOT is a sort of autonomous robot that is capable of navigating between machine cells. An autonomous robot, sometimes referred to as an autobot or autoboot, is a robot that exhibits a high degree of autonomy when performing behaviors or tasks. According to the usual uses of COBOTS, they may be used in the following areas:

- Logistics
- E-Commerce
- Fulfillment of orders
- Transportation and sorting of raw materials
- Sorting of parcels
- Inventory management
- Warehousing
- R&D
- Health care
- Manufacturing
- Biotechnology

6.5.2   SUSTAINABILITY ANALYSIS

Science and technology evolve together. We are in the early stages of a new field. Proven competence in difficult environments and robot survivability can influence COBOT adaption. Environment friendly robot safety and reusability are included in the evaluation process for sustainability. In dangerous situations, the robot should work well while utilizing little energy. To determine mechanical or computational costs, a sustainable system uses the battery consumption rate. It saves energy without sacrificing performance. To stabilize and maneuver a robot requires a lot of energy. In the workplace, robots must adapt to their surroundings and avoid clashing. Unexpected incidents and hazardous materials can damage a COBOT in tough terrain. Fire damage is a possible source of harm in harsh settings. Visible obstacles can affect in hostile situations. The invisible barrier distance must be constantly monitored. The sustainability study and its consequences are discussed in the next sections. Customers must be happy for future uses of this technology to succeed. So, the robot's social and environmental consequences are missing from the sustainability evaluation.

6.5.3   ENVIRONMENTAL IMPACT

The COBOT was constructed entirely of stainless steel. Steel is an environmentally friendly material. Due to its non-toxic coating, it is 100% recyclable and does not produce harmful run-off. A simple switch to stainless steel from non-recyclable materials can have a noticeable impact. Stainless steel is manufactured from scrap metal, with up to 70% of the material recycled. Due to advancements in process technology, stainless steel manufacturing consumes less energy. If stainless steel is not recycled, it has no adverse effect on the land or groundwater. When compared to other materials, it has a minimal environmental impact, and its impact reduces as it is used and recycled. Secondly, instead of fossil fuels, batteries are used to power and operate COBOTS. Carbon dioxide and other greenhouse gases are released into the atmosphere when fossil fuels are burned, resulting in global warming and climate change.

6.5.4 HAZARD IDENTIFICATION AND SAFETY MEASURES

These dangers are classified as follows:

1. Inspect for malfunctioning sensors; sensor functionality is important to the operation's success. A bare electric line has the potential to electrocute. Electrical wiring must be insulated in accordance with ISO standards for mobile robots.
2. Additional caution should be exercised during prototype dry runs and testing to avoid operating the COBOT in a dynamic environment. It has the potential to cause damage to the COBOT's components.
3. Improper handling of materials during manufacturing might jeopardize the team's health.
4. The sensor should be positioned in an area with an ambient temperature; if it is put in an area that is too hot, it will lose accuracy.
5. During testing, the COBOT should not be permitted to exceed soft limitations, since this may cause damage to the components.
6. Charge the battery in an area that is well-insulated. When working with lead acid batteries, considerable caution and the application of the proper voltage are required. Connect the appropriate switches during charging. A simple approach would be to create a generic SOP booklet and post it in the plant.

6.6 PATH PLANNING OF COBOT

The COBOT is designed to move the objects from one machine to another machine following the shortest path. Various algorithms are used to follow the shortest path, but the best algorithm is A* algorithm. A* algorithm is one of the leading and prevalent strategy utilized in pathfinding graph traversals. It works based on the following equation [6.13]

$$F = G + H \tag{6.33}$$

Where G is the cost of each cell from the start node. Where H is the heuristic value of each node from the goal point. The result of the A* algorithm is shown below

In Figure 6.6, white blocks are the obstacle which will be avoided while selecting the shortest path. It searches all the nearest cell to find the shortest path until it reaches its goal position.

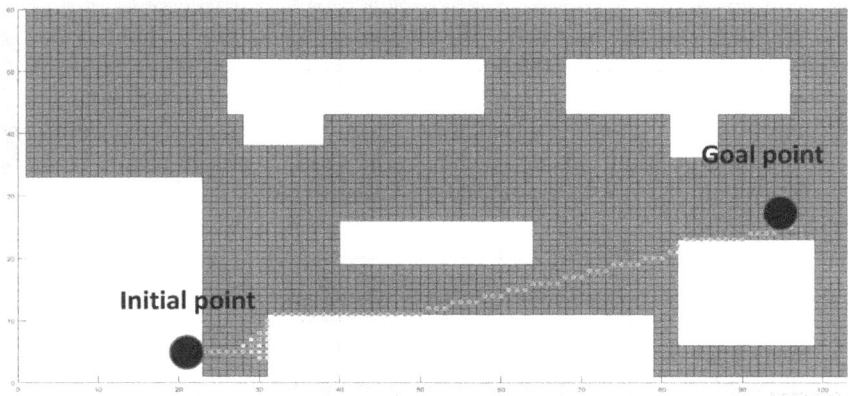

FIGURE 6.6   A* Algorithm Based Shortest Path Calculation by Avoiding the Static Obstacles.

6.7   CONCLUSION

In this chapter, the collaborative robot is designed to pick and place the objects from one node to the other node. CAD modeling of the COBOT is done by using the modeling software such as Solid Works. After the CAD modelling different software's are used for analysis such as Ansys are used to calculate the stress and displacement analysis. MSc Adams is used to calculate the torque of each motor and validated by the analytical analysis. Weight property index is used for material selection. After the analysis, mathematical modelling is performed to calculate the velocity of each wheel based on the desired velocity of the robot. Path planning is done to find the shortest path by avoiding the obstacles. A* algorithm is chosen among others path finding algorithm to find the shortest path.

In future, the robotic arm will be installed over the COBOT which will be used to pick and place the objects. Moreover, A LIDAR will be installed on COBOT to avoid the dynamic obstacles by using velocity obstacles method and machine learning techniques.

REFERENCES

[1] S. Kirsner. (2012, March 19). *Amzaon buys warehouse robotics start-up Kiva systems for $775million.* Available: http://archive.boston.com/business/technology/inno-eco/2012/03/amazon_buys_warehouse_robotics.html
[2] S. Kirsner. (2013, December 1). *Acquaisation puts amazon rivals in awkward spot.* Available: www.bostonglobe.com/business/2013/12/01/will-amazon-owned-robot-maker-sell-tailer-rivals/FON7bVNKvfzS2sHnBHzfLM/story.html
[3] Wecon Systems. (2019). *Introducing the OTTO 100 and OTTO 1500.* Available: https://weconsystems.com/introducing-the-otto-100-and-otto-1500/
[4] *Otto Motors.* Available: https://ottomotors.com/
[5] EasyRobotics. *Easy work (former ER 5)—mobile cobot platform.* Available: https://easyrobotics.biz/product/er-5-easywork/

[6] B. Heater. (2018, May 26). *Kiwi's robots deliver food to hungry Berkeley students.* Available: https://techcrunch.com/video-article/kiwis-robots-deliver-food-to-hungry-berkeley-students/

[7] W. Kane. (2018, May 31). *Those four-wheeled robots on campus, explained.* Available: https://news.berkeley.edu/2018/05/31/those-four-wheeled-robots-on-campus-explained/

[8] B. Berman. (2019, November 7). *Burrito delivered by bot, as long as students don't trap it.* Available: www.nytimes.com/2019/11/07/business/kiwibot-delivery-bots-drones.html

[9] (2018, November 26). *Juan miguel hernández bonilla four colombians, among the most innovative in Latin America.* Available: www.elespectador.com/noticias/ciencia/cuatro-colombianos-entre-los-masinnovadores-de-latinoamerica/

[10] N. Firth. (2017, December 8). *Food delivery robots are teaching themselves how to cross roads.* Available: www.newscientist.com/article/2155830-food-delivery-robots-are-teaching-themselves-how-to-cross-roads/

[11] *Kiwibot.* Available: www.kiwibot.com/

[12] E. W. Dijkstra. "A note on two problems in connexion with graphs," *Numeric Mathematika,* vol. 1, no. 1, pp. 269–271, 1959.

[13] P. E. Hart, N. J. Nilsson, and B. Raphael. "A formal basis for the heuristic determination of minimum cost paths," *IEEE Transactions on System Science and Cybernetics,* vol. 4, no. 2, pp. 100–107, 1968.

[14] E. N. Moret. (2003). "Dynamic modeling and control of a car-like robot," *Virginia Tech electronic thesis and dissertations.* Available: http://hdl.handle.net/10919/31535

[15] W. Ren, Q. Gu, D.-X. He, and J. Zhao. "Modelling and implementation of a car-like mobile robot for trajectory-tracking," *International Journal of Modelling, Identification and Control,* vol. 19, no. 2, pp. 150–160, 2013.

[16] E. Maulana, M. A. Muslim, and V. Hendrayawan. "Inverse kinematic implementation of four-wheels mecanum drive mobile robot using stepper motors," in *2015 international seminar on intelligent technology and its applications (ISITIA).* Piscataway, NJ: IEEE, 2015, pp. 51–56.

[17] N. Seshadri, S. N. Pramod, S. Vidat, P. Shashank, and P. K. Kumar. "Selection of optimum design among different vane designs by weighted average method," *Journal of Mechanical Engineering,* vol. 11, no. 2, 2021.

[18] *Maximum von mises stress criterion.* Available: https://help.solidworks.com/2013/english/solidworks/cworks/r_maximum_von_mises_stress_criterion.htm

7 Design, Manufacturing and Sustainability Analysis of Textile Spinning Machine

G. Hussain, Wasim A. Khan, K. Rehman,
G. Abbas, Muhammad S. Khan,
Muhammad S. Qaiser, Muhammad U. Kaimkhani,
Sayed Qaisar Hussain, Syed Essa Rasan and
Ulfat Hussain

CONTENTS

DOI: 10.1201/9781003220985-7

Nomenclature

μ_{RT}:	Coefficient of friction between the ring and the traveler
μ_{YT}:	Coefficient of friction between the yarn and the traveler
β_1:	Angle of wrap of yarn at guide eye
β_2:	Angle of wrap of yarn at the traveler
R:	Ring radius
ω:	Angular velocity of traveler
M:	Traveler mass
H:	Balloon height
T_{B2}:	Yarn balloon tension above the traveler
R_B:	Bobbin radius
α:	Angle of lead of the yarn, defined as sin α: RB/R
ρ_A:	Density of air
C_D:	Air drag coefficient
r_{max}:	Maximum radius of the balloon, radius of the balloon is taken as r
eff:	efficiency
Ne:	English count
L:	Delivery speed in m/m

7.1   INTRODUCTION

The spinning machine is the fundamental machine in the textile industry. The spinning machine turns loose fiber into firm threads by imparting a twist. This machine has developed throughout human history to cater to changing demands. Likewise, engineers face a new demand from machines in today's world. The most critical aspect of modern-day manufacturing is the environmental effects. The machines designed today must have maximum sustainability considering the depletion of resources. The machine must be environmentally friendly in terms of carbon emissions to tackle global warming. Therefore, this spinning machine is designed on the lines of prioritizing sustainability needs and environmental impacts along with facilitating small and medium scale enterprises.

The manufacturing process uses the principle of functional reverse engineering by the sourcing approach implemented. The weighing arm and top rollers, due to the complexity of design and manufacturing constraints, were sourced ready-made. The rest of the parts were manufactured considering the dimensions of the sourced parts.

Thus, the principles of functional reverse engineering were implemented which resulted in a lower manufacturing cost and time.

7.2   SCOPE OF THE WORK

This chapter will cover the design, manufacturing, and electrical interfacing of a small ring spinning machine. The machine is designed to facilitate small and medium scale enterprises. An industrial-scale machine on average has eight hundred spindles, however, this prototype will have two spindles. The sustainability section presents sustainability analysis in the manufacturing as well as operation of the machine. Environmental friendliness and sustainability were considered significant parameters during the design and development phase of the machine. Since the target market is SMEs and developing economies, it was a design consideration to minimize costs.

7.3   LITERATURE REVIEW AND RESEARCH GAP

Ring spinning is a commonly employed yarn spinning operation that uses a ring spinning frame that drafts roving (fibers prepared into loose strands ready for spinning where the fibers have been laid parallel to some extent), twists the yarn, and winds it on the bobbin, continuously and simultaneously in one operation. One of the benefits associated with the ring-spinning apparatus is that comparatively a greater proportion of fibers are aligned in a parallel manner [1].

The ring spinning frame was invented by the American John Thorp in 1828. In the preface to his 1987 book, Klein estimated that there were at that time 160 million ring spindles in operation throughout the world. However, since its inception, there hasn't been a major modification in the drafting process since the quality of the yarn surges not more than 1% before the cost outweighs the benefit [2].

Among the most impressionable factors, the spinning geometry is significant for the process since it affects the end breaks, tension conditions, and generation of fly, yarn hairiness, and yarn structure. Spinning geometry constitutes all the distances, angles, and inclinations cumulatively present between the different elements of the assembly.

Another major drawback concerned with the mechanism is the friction between the ring and traveler. The heat generated because of this friction majorly deteriorates the productivity of the process. Magnetic bearing systems are being used nowadays to prevent contact and eliminate friction.

A spinning triangle is a major detriment to the strength of the yarn. At the exodus from the front rollers, a triangular bundle of fibers without a twist is formed at one end. This is formed because the width of the fiber bundle emerging from the drafting system is many times the diameter of the yarn to be spun. This provides a vulnerable point at which ends may break. This cannot be entirely prevented but can be significantly minimalized by keeping the twist speed high.

Significant improvement has been seen in ring-spinning technology in recent years. By the application of optical sensors, the roving bobbin transfer has been optimized such that once the bobbin has been completely wound, it is replaced with an empty bobbin without any wastage of time and manual labor [3].

In recent years, in terms of efficiency, production, and quality of yarn, large numbers of improvements occurred in ring-spinning such as in air pipes, air suctions, and air conditioning requirements due to these changes the advantages are taking in terms of fine yarn count, high strength yarn and excellent in yarn evenness.

Additionally, there have been improvements like increasing the spindle speed up to 20,000 rpm, roving feed stop motion, and automatic creel of the bobbin. This has significantly optimized the process and enhanced process efficiency.

Based on the above literature analysis, the following key points can be deduced as key guidelines for optimum operational proficiency of the spinning machine:

- Ring spinning apparatus is the most used design for spinning yarn at a small scale.
- A major change in the original design is not beneficial since it does not enhance the quality or yield of the yarn before the cost of the upgrade outweighs its benefit.
- The affinity of the machine elements in the assembly contributes towards the enhancement of the yarn quality and strength. Hence, the assembly needs to be compact.
- The friction among the machine parts, and between the thread and the machine parts hampers the productivity of the process. Hence, friction needs to be minimal.
- A spinning triangle is formed at the ends of the thread. This provides a site for wear to settle in. This can be minimalized by keeping the twist speed high.
- The size of the yarn package is limited by the ring diameter, which must be small to increase the spindle rotation at the same traveler speed.
- The optimization is mainly focused on the cost aspect.
- There is a greater emphasis on making the design sustainable and environment-friendly in the new design and manufacturing process.
- The newer design will use DC motors instead of the induction motors used typically that are 30% more efficient.

The previous design of ring spinning machines is focused on maximum output without proper consideration of the sustainability aspect. The designs mainly focus on improving yields at the expense of environmental pollution in terms of carbon emissions. The existing spinning machines are designed to be feasible for large-scale industries that produce the product at massive scales. The size of the machine makes it difficult to keep the operations sustainable since it required in-house generators that involve fossil fuels.

For the reference of the reader, Figure 7.1 shows the final machine as a product with the drafting and spinning assembly.

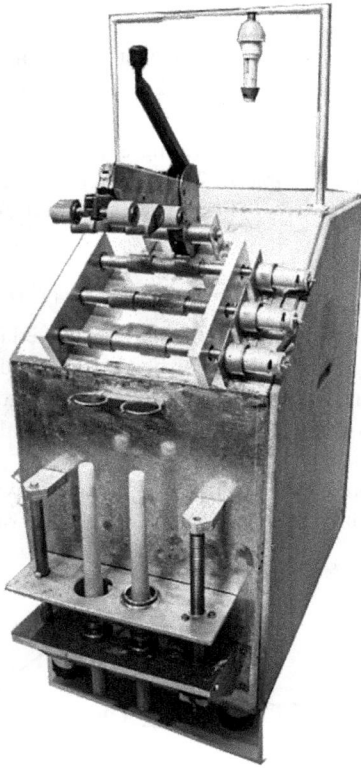

FIGURE 7.1 Photograph of Spinning Machine as a Product.

7.4 OBJECTIVES

The main objective of this book chapter is to explain the design methodology followed in the manufacturing of ring spinning machine. The areas of interest are to execute an efficient sourcing strategy to save costs and analyze the sustainability and environmental impacts of manufacturing and operating the machine.

7.5 DESIGN AND ANALYSIS

The assembly is composed of three main components: drafting system, spinning system, and frame. The technical aspects of the component design, working principle, and inter-working of components are further discussed. Dimensional accuracy was deemed as a necessary parameter for the initial design phase. Initially, a rough design was drafted with approximated dimensions to commence the process. After consultation of different standards and technical literature. The dimensions of the components were finalized. These dimensions still were subject to modifications based on further alterations in the design.

7.5.1 Governing Equations

The schematic in Figure 7.2 shows a spindle as it rotates around its axis. The bobbin is in its place. The yarn descends from the guiding eye towards the ring forming a balloon-like shape.

The height of this balloon from the guiding eye to the position where yarn sticks to the bobbin is given by the variable H. The radius of the ring is given by R, and the radius of the bobbin is given by R_B. These two radii are used in the formula below:

$$\sin \alpha = \frac{R_B}{R} \tag{7.1}$$

Equation 7.1 is used in other formulae which are displayed on the subsequent pages. Furthermore, is the radius of the balloon formed by the circular motion of the yarn as it comes down from the guiding eye. In later formulae, usage of r_{max} is common; this variable refers to the maximum radius achieved by the balloon due to the circular motion of the yarn.

The traveler rotates around the bobbin while in contact with the ring. This traveler has a certain mass M and it rotates at a certain velocity ω depending on the rotation of the spindle.

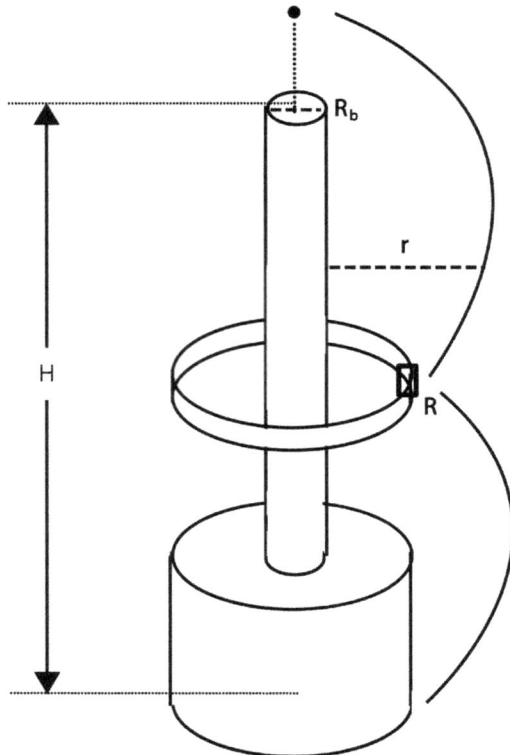

FIGURE 7.2 Schematic of Spinning Machine Spindle.

Now with the above variables defined, the equation used to calculate the tension in the yarn balloon right above the traveler is formed as Equation 7.2.

$$T_{B2} = \frac{a * \mu_{RT} * M * R * \omega^2}{\sin \alpha} \tag{7.2}$$

Where 'a' is a constant ranging from 0.55 to 0.62.

Furthermore, the calculation of air drag on the yarn is necessary so some more variables and constant need to be calculated. The density of the air is given by ρ_A, and the drag coefficient is by C_D. The drag coefficient is modeled for a cylindrical-shaped object as the yarn can be thought of as a cylinder with a fixed diameter d_y and a certain length. After plugging in the other variables in the equation, the horizontal component of air drag becomes:

$$F_D = \frac{\rho_A C_D d_y \omega^2 H r_{max}^3}{4R} \tag{7.3}$$

Figure 7.3 shows the different forces and their respective directions on the cross-section: F_D is the horizontal drag component of yarn balloon tension above traveler, T_{B2}:

$$F_D = T_{B2} \cos \theta_2 \tag{7.4}$$

After rearranging the above formula, we get the equation of angle of yarn in the balloon with respect to the x-axis:

$$\theta_2 = \cos^{-1} \frac{F_D}{T_{B2}} \tag{7.5}$$

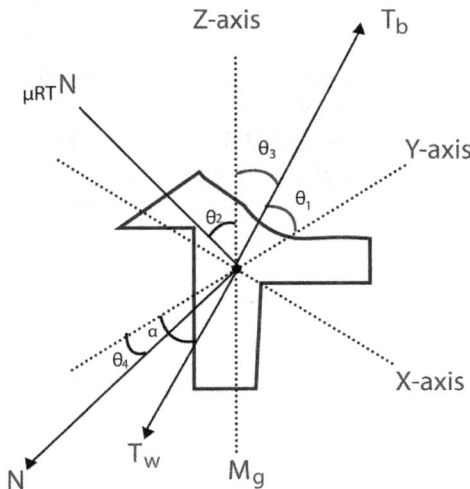

FIGURE 7.3   Forces Acting on Traveler's Cross Section during Spinning.

After a good degree of approximation, the angle of yarn in the balloon with respect to the z-axis is calculated with:

$$\theta_3 = \cos^{-1}\left(sin\theta_2\right) \tag{7.6}$$

To find another important angle, the angle of inclination of the traveler with respect to the y-axis, another formula is used. The formula is stated below:

$$\theta_4 = \sin^{-1}\left[\frac{\alpha\,\mu_{RT}\,(T_{B2}\cos\theta_3 - Mg)}{(T_{B2}\sin\alpha - a\cos\theta_2)}\right] \tag{7.7}$$

Where g is gravitational acceleration = 9.81 m/s².

After finding out all the variables above, a few more are needed as well. The friction coefficient between the yarn and traveler is μ_{YT}. The friction coefficient between yarn and guide eye is given by μ_{YL}. The angle of wrap of the yarn around the traveler is given by:

$$\beta_2 = -\frac{\ln a}{\mu_{YT}} \tag{7.8}$$

Where a is a constant between 0.55 and 0.62.

The value of β_1 fluctuates from zero to 50 degrees.

Now after all these variables and constants are determined, the equation of spinning tension is derived. Equation 7.9 is used to find whether yarn will break or stay intact during the process of spinning. The formula of spinning tension is:

$$T_s = \frac{\dfrac{\mu_{RT}MR\omega^2 + F_D\cos\theta_4}{e^{\mu_{YT}\beta_2}(\sin\alpha\,\cos\theta_4 + \mu_{RT}\,\cos\alpha)}}{\mu_{YL}\beta_1} \tag{7.9}$$

In the book, Advances in yarn spinning technology the tensions calculated from the formula above range from 20–25 cN.

The spindle on average rotates at 10,000 to 12,000 rpm. This is achieved with a 1,500W 2,000 RPM motor. A gear ratio of 3:1 increases the RPM from 2,000 to 6,000 and then the pulley ratio to the spindle is half so that gets an RPM of 12,000.

Production of yarn is given by Equation 7.10:

$$P = \frac{L*1.0936*60*eff}{Ne*36*840*2.2045} \tag{7.10}$$

The constant of 20.2045 is present in the equation to convert the production mass from pounds to kilograms.

7.5.2 Mechanical Design

The ring spinning machine assembly consists of two major subassemblies:

- Drafting system subassembly
- Spinning system subassembly

2. *Drafting System*

The drafting arrangement is the most important part of the machine. It is a design and fabrication process that controls the evenness and strength of the thread. As a result, the bulk and weight per unit length of semi-processed textile material are reduced and the fibrous components of the material are parallelized simultaneously.

The drafting mechanism to be utilized in our ring spinning machine is known as the three-over-three double apron drafting system. This system consists of two sets of three rollers: top rollers, and bottom rollers. The rollers rotate with the help of ball bearings. The top rollers exert a force on the bottom rollers (shafts) as they rotate. As the yarn passes through the rollers, it is pressed due to force from the top rollers which cause its dimensions to change.

The drafting mechanism is divided into two zones. The first zone or back zone is called break draft and the second or front zone is designated as the main draft. A pair of aprons is connected to the middle rollers and is made to move at their surface speed. The aprons hold the yarn fibers and assist in keeping them moving at the surface speed of the middle rollers while preventing the short fibers from being dragged forward by those fibers that are being nipped and accelerated by the front rollers.

3. *Top Rollers*

Top roller quality is of major importance for spinning quality yarn. The smooth running of the top roller in direct contact with the roving influences the drafting result and herewith the quality. The first step in designing the top rollers was to create a metallic frame to be used for each roller.

The 18.9 mm diameter sections of the roller frame are to be coated with a rubber cot. The rubber cots selected for the rollers in our machine are high-quality BERKOL cots. This is because they have a long lifetime and high grindability, thus providing consistent yarn quality and reduced lap formation and yarn breaks. They also provide excellent yarn guidance which leads to the high efficiency of the ring-spinning machine [4].

The blue-highlighted portion of the roller represents the rubber cots. These cots, along with the advantages, also help protect the bottom rollers. At the ends of the roller, beneath each rubber cot, are two ball bearings to help rotate the roller. Each top roller has a total of 4 ball bearings.

Bottom rollers play a major role in ensuring yarn quality. Precise concentricity guarantees high yarn quality (evenness and strength). The running conditions determine the occurrence of yarn breaks. Precise running bottom rollers allow extending the spinning limits into finer yarn count ranges. The bottom rollers are shafts with a circular cross-section. These rotating shafts are responsible for the drafting process. Each shaft has two sections that are engraved with helical grooves of length 30mm along the length of the shaft. These sections are 50 mm apart on the shaft. The pitch of the helical grooves is 1 mm, the helix angle is 8°

and the number of teeth is 50. The shaft usually rotates at a speed of 5,000 rpm to 10,000 rpm [5]. However, the actual rotational speed depends on the required diameter and quality of the thread.

A motor is connected to one end of each shaft to rotate it with the help of a needle ball bearing. As the shaft rotates, it draws the thread of cotton from the unspun cotton roving bobbin. The total length of the shaft is 205 mm, including the lengths of the grooved sections.

4   Spinning System

After passing through the drafting system, the processed, refined thread must be twisted and spun around a bobbin that is mounted on a spindle. This process is carried out by the spinning system which consists of three important components:

- Ring
- Traveler
- Spindle
- Bobbin

The ring travels up and down the spindle in vertical motion while the traveler, attached to the ring, simultaneously rotates around it, hence, winding the thread around the bobbin. The spindle and in turn, the bobbin, rotate in the opposite direction to the traveler. This causes the thread to twist while winding around the bobbin. The details of the ring, traveler, and spindle along with their CAD models are presented below.

7.5.2.1   Ring

The ring is one of the main components of a spinning machine. It controls the spinning process of the machine which is why the machine is called a ring spinning machine. The ring performs two important functions:

- Twists the thread to increase its strength
- Winds the thread around a bobbin through a circular traveler

The yarn is wound onto a cylindrical bobbin in the form of a balloon by the vertical motion of the ring around the spindle. The ring is mounted on a continuous ring rail using a cup. It has different shapes depending on the type of fiber used and the speeds of spinning. Rings come in different shapes and sizes. Some of the ring-types used in spinning machines are:

- T-flange ring
- Anti-wedge ring
- Cropped ring
- Inclined-flange ring

The most used ring is the T-flange ring. Since we are processing cotton using our machine, we will also be using this type of ring, keeping in mind the simplicity in its design. The traveler rotates along the T-flange of the ring. It comes in different diameters depending on the twist required and the speed of the spindle usually in the range of 42–58 mm [6]. The CAD model of the ring designed has a mean diameter of 47 mm and a four mm flange width.

7.5.2.2 Traveler

The traveler is a tiny, yet important component of the machine. It plays an important role in the twisting and winding of the thread on the bobbin. The traveler fits loosely over the upper flange of the ring. The flange and the ring act as a guiding track for the traveler and hence, under the pull of the yarn, the traveler moves about the circumferences of the ring. In the traditional setup, the yarn is threaded from the drafting rollers through a thread guide and the traveler and wound onto the bobbin. The weight of the traveler must be under the strength of the yarn wound on the bobbin to avoid excessive strain on the yarn and yarn breakage.

For spinning finer yarn count, round and C-travelers are commonly used as they provide less friction. We will be using C-traveler in our design as we are working with cotton. Modern spinning machines have very high speeds reaching 10,000 rpm. As the traveler rotates about the ring, friction is produced as the two surfaces rub against each other. The maximum speed at which the travelers can spin without burning due to friction is about 4,500 feet per minute, which correlates to a spindle speed of about 7,640 rpm [7].

7.5.2.3 Spindle

The spindle is the last and most important component in the spinning system of the ring-spinning machine. The spindle performs a variety of functions:

- Draws the yarn from the drafting system using centripetal force
- Rotates in the opposite direction to the traveler to twist the thread
- Holds the bobbin

FIGURE 7.4 Exploded View of the Ring, Cup, and Traveler Assembly.

The capacity of the machine is determined by the number of spindles that can be held in the spindle holder. The operating conditions/parameters of the spindle greatly affect the yarn quality. Poorly run spindles have adverse impacts on the yarn parameters. The spindle designed used by us can rotate up to a speed of 30,000 rpm using a motor One of the most important things to keep in mind is to align the centers of the spindle and the ring. The two components move relative to each other and can change position during each operation and thus, should be centered from time to time using mechanical or electrical devices. These balancing errors and eccentricity relative to the ring can significantly impact the quality of the yarn by interfering with the twisting and winding process of the thread around the bobbin. Spindles also control the energy consumption and noise level produced by the machine.

The complete spindle assembly consists of the following parts:

- Upper part
- Lower part
- Spindle clip
- Spindle washer
- Spindle needle bearing

The upper part of the spindle consists of the spindle blade and spindle wharf while the lower part consists of the spindle brake and bolster case.

7.5.2.1.1   Bobbin

The last and foremost component in the spinning system subassembly is the bobbin. The bobbin is mounted on the spindle and rotates due to the rotation of the spindle in the opposite direction to the traveler. The twisted yarn is wound around the bobbin to form a balloon of thread. Once the required maximum diameter (at the center) of the balloon is achieved, the bobbin is removed and replaced with a new one.

7.5.2.1.2   Working Principle of Ring Spinning Machine

The fiber goes through several processes before it is passed through the ring-spinning machine to get the refined final product. These processes include:

- Picking cotton fibers
- Carding
- Combing
- Drawing

These processes are used to prepare a form of fiber wound on a bobbin called the roving. The roving fiber is fed into the ring-spinning machine to be further drawn and then, spun and twisted around a bobbin to be used as a final product.

Roving (partially processed fiber), wound on roving bobbins, is passed through a series of drawing rollers that draw the strand of fiber to the desired final thickness. This process is called drafting. A bobbin, larger than the roving bobbin, is mounted on a spindle and rotates due to the rotation of the spindle at a constant speed. The yarn delivered from the drafting system is twisted as it wounds on the bobbin to form a balloon of thread. The twisting of the yarn depends on the rotational speed of the final pair of drawing rollers and the spindle. The traveler which moves around the bobbin on the balloon control ring, guides the yarn so that it is properly wound on the bobbin. The drag on the traveler causes the yarn to wind at the same rate as it is delivered by the final pair of rollers of the drafting system. Once the balloon of thread is formed on the bobbin, it is replaced by another bobbin [8].

7.5.2.4 Final Assembly

Figure 7.5 shows the CAD model of the ring spinning machine designed and manufactured for reference. The complete manufactured machine is shown in Figure 7.1.

7.5.3 ELECTRICAL DESIGN

The electrical design refers to the software-hardware interfacing of the ring-spinning machine. The interfacing is carried out to allow the operator to input the desired values of the machining parameters which include the rotational speed of the bottom rollers (shafts), twists per unit length, feed of the ring, etc. The machine will operate using a single stepper motor. However, using gear trains of different gear ratios, the speed of the shafts and the spindle can be kept different from one another. The interfacing will be carried out using a micro controller Arduino Uno.

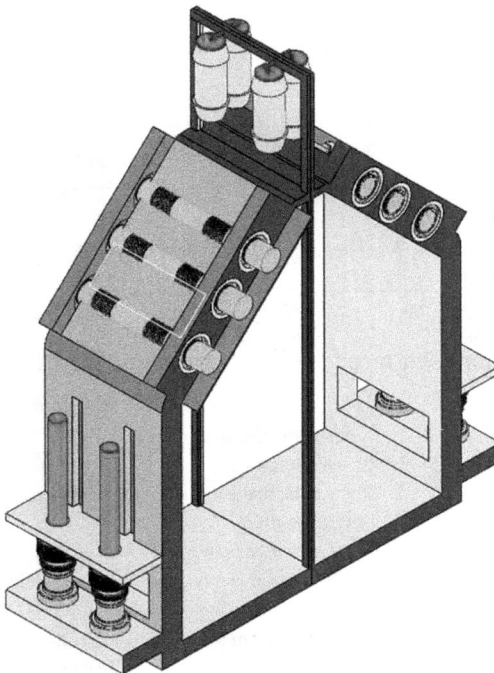

FIGURE 7.5 CAD Model of the Ring Spinning Machine with Weighing Arm Removed.

7.5.3.1 PWM Motor Speed Control Using Arduino

The next step in interfacing is to program the Arduino to control the speed of the motor which will, in turn, control the speed of the shafts. PWM or pulse width modulation is a technique that allows us to adjust the average value of the voltage that is going to the electronic device by turning on and off the power at a fast rate. This technique is used to control the motor speed in our ring spinning machine.

7.6 PHYSICAL MODEL DEVELOPMENT

The fabrication of the ring-spinning machine involved two steps: manufacturing or procurement of components and assembly of components. List of manufactured and outsourced components, process pan for manufactured components and details of outsourced parts is given in Table 7.1, Table 7.2 and Table 7.3 respectively.

The list of components used to put together a functional ring spinning machine are listed below:

TABLE 7.1
Shows a List of Manufactured and Outsourced Components.

Manufactured	Outsourced
Bottom rollers	Weighting arm
Top rollers	Rubber cots and apron
Moving table	Ring and traveler
Table shafts	Bobbin and hanger
Roller bearing housing	Spindle
Steel frame	Hanger
Supports	Balloon control ring

7.6.1 Process Plan for Manufactured Components

TABLE 7.2
Shows the Operations Required for the Manufacturing of Components.

Assembly	Part name	Operations for the manufacturing of the part
Drafting assembly	Bottom rollers	1. A cut shaft of 35 mm diameter was cut into three portions of length 250 mm. 2. Reduce the diameter to 20 mm from the ends and 30 mm at the central length of 50 mm. 3. Form a diamond pattern through the knurling process.
	Top rollers	1. Top rollers are identical to the bottom rollers and the same manufacturing techniques were employed.
	Roller rearing housing	1. Cut two 200 mm x 70 mm plates of thickness 12.5 mm. 2. Drill three holes into each steel plate with a 70 mm center-to-center distance. 3. Enlarge to 20 mm holes using a boring tool.

Assembly	Part name	Operations for the manufacturing of the part
Moving table assembly	Steel table	1. Cut two rectangular 300 mm x 100 mm steel plates from a steel sheet of thickness 5 mm.
		2. Drill 2 of the holes of 42 mm diameter in the central region for the spindles and one hole of 22 mm diamer at each end for the support shafts.
		3. Thread one hole of 22 mm diameter internally.
	Table shafts	1. Cut two 220 mm long steel shafts of 20 mm diameter.
		4. Produce external threading on the right table shaft corresponding to the internal threading of the steel table hole.

7.6.2 OUTSOURCED COMPONENTS

TABLE 7.3
Contains the Details of Outsourced Parts.

Outsourced part	Details
Weighting arm (Figure 7.6)	1. Contains the main drafting system that provides the drafting force.
	2. The distance between the in-line rollers is equal to the knurled cylinders in the bottom rollers.
	3. Distance between the shafts is equal to the distance between roller bearings.
Rubber cots	1. Protects the bottom rollers from wear and tear caused by friction.
	2. The thickness of the rubber cots is 10 mm.
	3. The middle top rollers have a rubber apron wrapped around them.
Ring and traveler	1. Traveler used is C-traveler type.
	2. The ring used has an inner diameter of 42 mm and an outer diameter of 54 mm.
Bobbin and hanger	1. Acts as feed to the drafting assembly.
	2. Made of plastic and are 200 mm long.
Spindles (Figure 7.7)	1. Made of aluminum.
	2. The total length of each spindle is 350 mm.
13. Balloon control ring	1. Guides the thread as it twists and winds on the bobbin to form a balloon.
	2. Made of aluminum.

FIGURE 7.6 Photograph of Weighting Arm with Top Rollers Attached.

FIGURE 7.7   Photograph of Spindle.

7.6.3   FINAL ASSEMBLY

The final assembly was executed in a series of steps:

- Welding of steel sheets to form the steel section of the machine frame.
- Installing 20 mm roller bearings in the roller bearing housing.
- Slotting the bottom rollers into the bearings.
- Joining the roller bearing housing to the frame using hexagonal nuts and bolts.
- Positioning the weighting arm on top of the bottom rollers and fastening it to the frame using an angle joint. This completes the assembly of the drafting system of the machine.
- Fastening the table shaft supports and the moving table assembly on the machine frame.
- Connecting the balloon control rings to the machine frame.
- Joining pentagonal wooden sides, using screws, to the steel section to complete the machine frame and the assembly of the Ring Spinning Machine.
- The final assembly is displayed in Figure 7.1.

7.7   IMPACT AND ECONOMIC ANALYSIS

7.7.1   SOCIAL SUSTAINABILITY

This research has direct implications in the textile industry. While keeping the target market strictly developing countries, it is found that a project of similar magnitude can effectively re-shape or reinforce the existing textile industry which is currently limited to heavy industry. Also, there is a visible discrepancy, since the readily available raw material is currently being utilized to a minimum extent; this can be attributed to the expenditure of heavy machinery required to produce a valuable final product.

The prototype can be utilized to create numerous manufacturing cells functioning at a smaller scale. This can decentralize the market, with more equal distribution of wealth and opportunities. State support for the growth and development of the small-scale industry is an important policy intervention aimed at addressing issues such as poverty and unemployment. Small scale industry, besides its several unique features such as frugal capital investment, resilience against recessions, ability to maintain growth over time, generation of agglomeration economies, and support for entrepreneurial talent, is primarily recognized for its potential to generate employment opportunities.

The impact of this project pertains mainly to the manufacturing of a profitable and functional machine, to substitute an otherwise expensive and behemoth option which although guarantees efficiency but lacks in practicality. In essence, the existing model is rescaled to bring the size, cost, weight, material usage, and functionality to a range where the machine could find application at a small scale or cottage industry scale level. The project provides an economical and practical remedy to a monopolistic industry that can facilitate SME's and developing countries.

7.7.2 ENVIRONMENTAL IMPACT

14 Carbon Emissions

The environmental impact of a project is determined by the number of carbon emissions produced throughout the development of the project and during its operation. The carbon emissions generated during the manufacturing processes were ignored. However, the carbon emissions due to materials and operation of the machine were undertaken.

15 Due to Electricity Consumption

Power rating - roller motor = 13 W

Power rating - stepper motor = 2.33 V×1.5 A= 3.5 W

Power rating - spindle motor = 12 V×5 A= 60 W

Total power consumption of the spinning machine = 13+13+13+3.5+60 = 102.5 W

Total energy consumed by the spinning machine in 1 hour/KWh = 0.1025 KWh

Kg CO_2/KWh produced for electricity consumption in US = 0.417305 Kg CO_2/kWh

Kg CO_2/KWh produced by a ring spinning machine unit = 0.417305× 0.1025 = 0.0428Kg CO_2/kWh

16 Due to Material

The main material used for the manufacturing of the ring-spinning machine is steel, as listed in below Table 7.2. Each kg of steel produces 1.9 kg CO_2. The estimation was made following this footprint.

TABLE 7.4
Shows the CO_2 Emissions for Different Parts Due to Material.

Parts	Mass in kg	CO_2 in kg/kg
Main body	5.5	10.45
Supporting blocks	2.2	4.18
Shafts	2.1	3.99
Moving table	1.2	2.28
Weighing arm	2.9	5.51
Spindle	1.1	2.09
Platform	2.4	4.56
Total	17.3	33.6

7.7.3 COST ANALYSIS

The machine cost is classified into three sub-categories:

7.7.3.1 Manufacturing Cost

Following is the breakdown of the total cost incurred in the manufacturing of the final product:

TABLE 7.5
Shows the Cost of Outsourced Components.

Cost	Amount in USD
Material cost	73
Machining cost	61
Outsourcing cost	140
Manufacturing cost	122
Total cost	396

7.7.3.2 Operational Cost

The main operational cost would concern with the electricity charges associated with the operation of the machinery. The following calculations were made to estimate the per hour electrical cost of operating the machine:

Power rating (roller motor) = 13 W
Power rating (stepper motor) = 2.33 V × 1.5 A = 3.5 W
Power rating (spindle motor) = 12 V × 5 A = 60 W

Total power consumption of the spinning machine/W = 13+13+13+3.5+60 = 102.5

Total energy consumed by the spinning machine in 1 hour/KWh = 0.1025 KWh

Average per KWh cost of electricity in US = 10.42 cents

Total electricity cost per unit hour of operation = 1.07 cents = 0.001 USD

7.7.3.3   Cost of Pollution

Cost of pollution implemented by government = 0.06 USD/kg CO2

Cost of pollution of operation for a single unit of Ring Spinning Machine 0.06×0.0389 = 0.0023 USD/hr

Cost of pollution due to material used = 33.6 kg CO2× 0.06 USD/kg CO_2 PKR/ kg CO2 =1.974 USD

7.7.4   COMPLETE COST BREAKDOWN

7.7.5   ONE-TIME COST

TABLE 7.6
Shows One-Time Costs.

Cost	Amount in USD
Manufacturing process	396
Pollution due to material	1.974
Total	398

7.7.6   HOURLY OPERATIONAL COST

TABLE 7.7
Shows the Hourly Operational Costs.

Cost	Amount in USD
Due to electricity per hour	0.001
Due to pollution per hour	0.0023
Total	0.0033

The initial cost is dominated by the manufacturing cost with negligible cost in pollution due to material. Therefore, it can be concluded that the ring-spinning machine has minimum CO_2 effects attributed to the extraction of material.

Hourly operational costs are 0.3 cents per hour approximately. This is a very low cost, and it shows that the machine is feasible for Small and Medium Enterprises and developing countries.

7.8   COST OPTIMIZATION

Cost of the machine = 396 USD
Cost of similar machine with 6 spindles = 11,000 USD (anytester model number AT206)
Cost per spindle = 1,833 USD
Cost of two spindles = 3,666 USD
% difference = 89.198% decrease

Hence, it can be concluded that the cost efficiency achieved through this design is approximately 89%.

7.9   CONCLUSION

This work entailed a thorough process of design and fabrication of a small-scale spinning machine considering sustainability. The new design has improved characteristics in the carbon emissions during operation. This is achieved by using DC motors that are more efficient than traditional induction motors. The operation cost of 0.003 USD per hour is a testimonial of the environment-friendliness of the new design.

The second main objective was the feasibility of the design for SMEs. The cost optimization analysis provided above yields an 89% decrease in manufacturing cost. This means that the machine is inexpensive. It is feasible for small-scale industries and developing economies that can add value to their raw material at a much lower cost. In conclusion, it is reasonable to state that the machine is environment-friendly, inexpensive, and sustainable. It fulfills the criteria of a machine developed for the modern-day.

REFERENCES

[1] H. A. Mckenna, J. W. S. Hearle, N O'hear, and Textile Institute, *Handbook of fibre rope technology*. Abington: Woodhead, 2004.
[2] W. B. Fraser, "On the theory of ring spinning," *Philosophical Transactions of the Royal Society of London. Series A: Physical and Engineering Sciences*, vol. 342, no. 1665, pp. 439–468, Feb. 1993, doi:10.1098/rsta.1993.0028.
[3] N. Bakhsh, M. Q. Khan, A. Ahmad, and T. Hassan, "Recent advancements in cotton spinning," *Textile Science and Clothing Technology*, pp. 143–164, 2020, doi:10.1007/978-981-15-9169-3_8.
[4] Reiter, "BERKOL® top rollers for high yarn quality," *www.rieter.com*. www.rieter.com/products/components/ring-and-compact-spinning/cots-berkol (accessed May 26, 2022).
[5] K. Fujisawa, "An approach to super high-speed ring spinning frame," *Journal of the Textile Machinery Society of Japan*, vol. 17, no. 1, pp. 1–9, 1971, doi:10.4188/jte1955.17.1.
[6] L. N. Periaswamy, J. C. Balasubramaniam, P. Nagarajan, K. K. Balaraman, R. Duraisamy, and S. K. Ramachandran, "A ring spinning and twisting machine," EP1095178B1, European Patents. Dec. 21, 2005.

[7] S. S. Bhattacharya, *Value added textile yarns-manufacturing techniques and its uses value added textile yarns-manufacturing techniques and its uses.* New Delhi: Om SaiTech Books India, 2020.

[8] Truents, "Standard staple yarn spinning procedures," *Textile School*, Jun. 21, 2011. www.textileschool.com/134/standard-staple-yarn-spinning-procedures/

8 An Introduction to Additive Manufacturing (AM) of Metals and Its Trends in Wire-Arc Additive Manufacturing (WAAM) Simulations

Rameez Israr and Johannes Buhl

CONTENTS

8.1 INTRODUCTION

8.1.1 ADDITIVE MANUFACTURING (AM)

Additive manufacturing (AM) which is also known as additive fabrication, three-dimensional printing (3DP), and layer manufacturing is a process for producing objects layer over layer [1]. The first AM method for metals was invented in the 1980s, and it is currently widely utilized in the manufacturing of components and

DOI: 10.1201/9781003220985-8

parts including industries, e.g., automobiles, aviation, marine and civil. Laser metal deposition (LMD) [2], three-dimensional printing (3DP) [3], laminated object manufacturing (LOM) [4], stereolithography (SLG) [5], and selective laser sintering (SLS) [6] are some of the different types of AM techniques. AM enables the direct production of complex 3D parts of various shapes from its computer aided design (CAD) model using computer-controlled movements of the same machine, with no need for specialized tools for performing a specific operation, such as grinding tools, milling tools, forming tools and forging tools. Consequently, production time is reduced, and the process becomes cost-effective even for unit or small-batch productions [7] as shown in Figure 8.1.

In the past, various researchers highlighted the advantages of AM products concerning their application, strength, productivity and cost. Guo et al. [7] elaborated the effectiveness of AM processes in various fields of science including automotive, aerospace, power, biomedical and power. Lim et al. [8] focused on large-scale additive manufacturing components utilized in the construction and architectural sectors. They claimed that using the AM method to manufacture concrete provides them with design flexibility to produce any shape precisely and thereby reduces the workforce. Baufeld et al. [9] used the AM technique to create a variety of products and then investigated their microstructure and mechanical characteristics. The mechanical properties of AM components were equivalent to those manufactured by conventional methods. Brandl et al. [10] compared the mechanical properties of components made with two different AM methods: laser-beam AM and wire-arc AM. They discovered that components created using AM methods are strong enough for aeronautical purposes. Jiang et al. [11] described the AM process as an ecofriendly technique that can significantly minimize material consumption, material waste, production time, and energy consumption. Camacho et al. [12] argued that AM methods have the ability to minimize labor costs and material waste despite achieving customized complicated shapes. Based on the facts presented above, AM

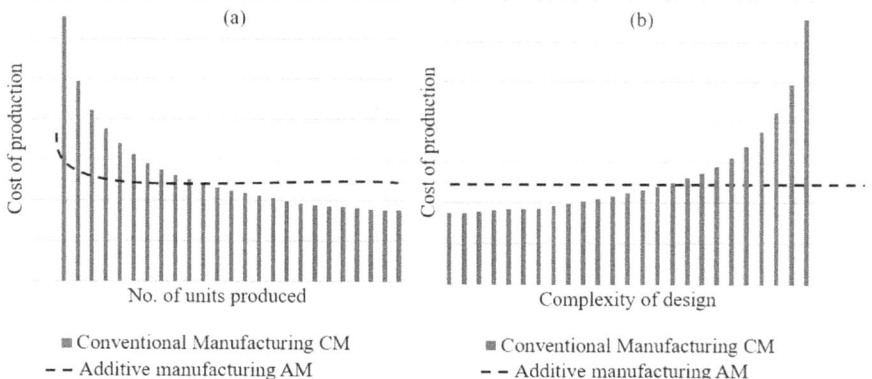

FIGURE 8.1 A Comparison of the Cost of Producing a Component (a) and Design Complexity (b) in Additive versus Traditional Manufacturing.

can be considered as one of the most effective strategies for increasing production and reducing wastage.

For the production of metal parts, several AM techniques exist. These include shape deposition manufacturing, SLS, electron beam freeform fabrication, and direct energy deposition (DED) [13]. These production methods are generally divided into two groups [14]:

- Powder bed fusion
- Direct energy deposition

8.1.2 POWDER BED FUSION PROCESS

In the powder-bed fusion process, the metal powder is melted in a predefined pattern using the energy from the heat source. Commonly used heat sources include laser beam and electron beam. The metal powder is stored in a container having a retractable piston and a leveling system as shown in Figure 8.2. A heat source is focused on the top layer of metal powder and melts it along a predetermined pattern. After forming a metal layer, the powder bed drops down one step. A new powder layer is then distributed to build a second layer. Scanning is done layer after layer until the entire part is generated.

Because this technique can only be carried out in a controlled container with high-quality powder, a large quantity of inert gas and costly machines, it is more

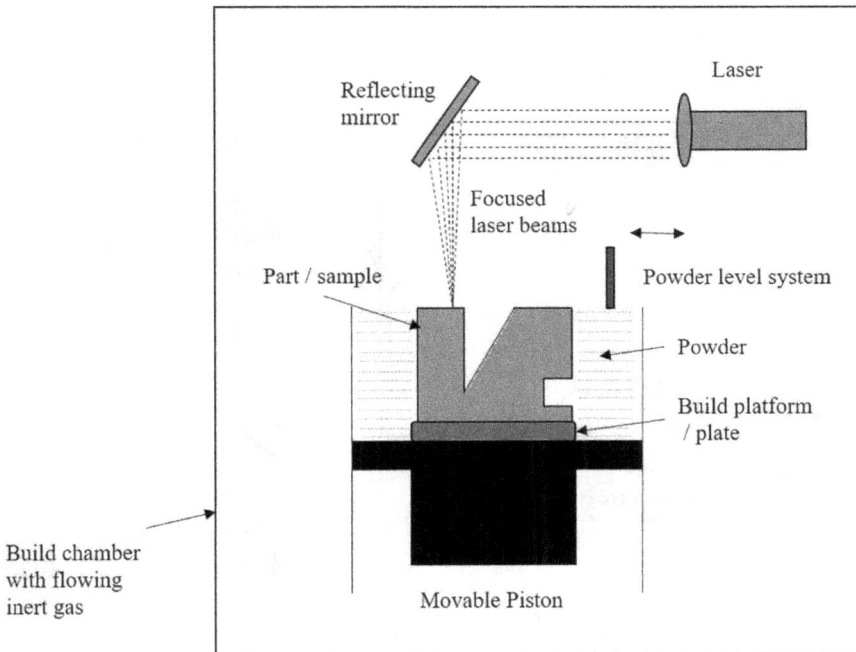

FIGURE 8.2 Description of Laser-Based Powder-Bed Fusion Process.

expensive than other procedures and is only suitable for complex and expensive small to medium-sized parts such as medical instruments, gems, and smaller engine components [15].

8.1.3 DIRECT ENERGY DEPOSITION PROCESSES

In DED, the metal powder or wire are used as feedstock material, where they are melted through either a laser or an electric arc and deposited on the substrate. As the deposition rate is higher and there is no requirement for a closed container, DED can produce medium to large-sized components [14]. Figure 8.3 shows a schematic depiction of a laser-based DED method. The DED technique that uses metal wire is often referred to as the wire-based deposition process.

8.1.4 WIRE-BASED DEPOSITION PROCESS

In wire-based deposition processes, metal wire is used as a filler material along with an energy source. The wire-based deposition techniques could be classified into three groups based on the energy source i.e., laser beam, electron beam and electric arc [13]. Because of its higher wire-consumption efficiency, availability, high deposition rates and low price, an electric arc is considered the most efficient heat source [16,17]. Table 8.1 presents a comparison of deposition rates and wire-consumption efficiency for several wire-based deposition methods. As can be observed, the deposition rate of electric arc-based deposition/welding is around 20–25 times that of the

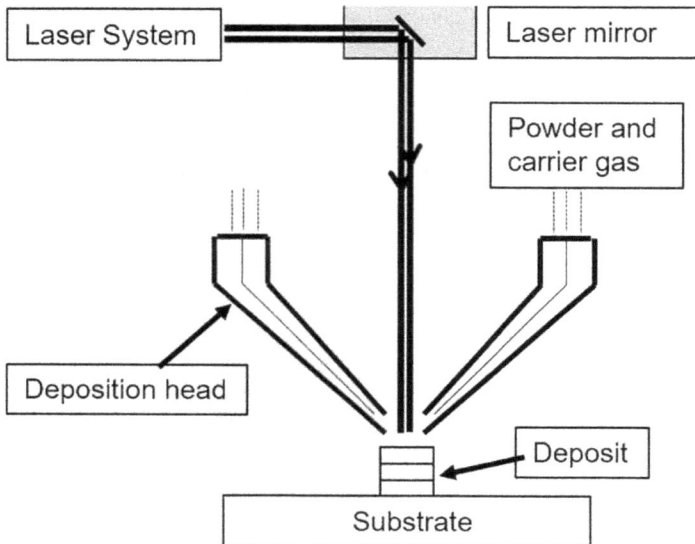

FIGURE 8.3 Schematic Depiction of Laser-Based Direct Energy Deposition Process.

TABLE 8.8

Deposition Rate and Process Efficiency of Different Wire-Based Deposition Methods [17].

Process	Deposition rate	Wire-consumption efficiency	Setup cost
Electron-beam based deposition	2–10 grams/min	15–20 %	High
Electric-arc based deposition	50–130 grams/min	90 %	Low
Laser based deposition	2–10 grams/min	2–5 %	High

other two methods. As a result, the wire-arc welding technique can be utilized to efficiently manufacture a variety of metal parts and components.

8.1.5 WIRE AND ARC ADDITIVE MANUFACTURING (WAAM) PROCESS

The evolution of AM has resulted in the wire and arc additive manufacturing (WAAM) technique, which uses a traditional wire-arc welding process to produce medium to large-scaled 3D components [18] for a wide range of materials including steel, titanium alloys, nickel alloys, aluminum, tungsten and tantalum [19]. The WAAM process is known for its versatility and productivity without sacrificing welding quality [20]. Due to its low buy-to-fly ratio of up to one, WAAM is regarded as a cost-effective method. Hence, as illustrated in Figure 8.1, the cost of an individual item decreases while process efficiency enhances [21,22].

In the WAAM process, the computerized control robotic arm with attached filler metal wire and welding torch regulate the arc welding process [23]. The electric arc melts the filler metal wire and the molten metal is deposited layer over layer on the substrate until the whole component has been produced with very little to no post-processing requirements [24]. Figure 8.4 depicts a schematic diagram of a typical WAAM process. The welding torch, which has a filler metal wire attached to it, can be seen generating an electric arc and melting the metal wire. Shielding gas surrounds the welding process, protecting the welding layers from contaminants and reducing heat accumulation. The movement of the welding torch is from left to right. As the layers are laid on the substrate, the welding torch travels upward to weld the next layer above the previous layer. The procedure repeats until all of the layers have been printed to form a wall-like structure.

Different welding techniques can be adapted to carry out the WAAM process, i.e., gas tungsten arc welding (GTAW), plasma arc welding (PAW), and gas metal arc welding (GMAW). In the GTAW and PAW, a non-consuming electrode produces an arc for welding the part, whereas, in the GMAW, an electric arc melts the metal wire and the molten metal transfers to the substrate. GMAW is a preferable option for WAAM processes due to its high deposition rate, low equipment cost and superior structural integrity [23]. Despite these enormous advantages, WAAM parts are subjected to significant distortions and residual stresses. Numerical simulation is therefore an efficient way of mitigating these shortcomings [27].

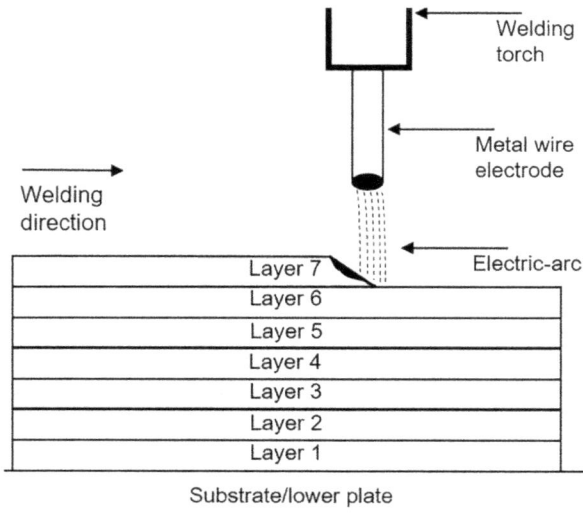

FIGURE 8.4   Illustration of WAAM Process.

8.2   SIMULATION METHODS FOR WAAM

The finite element method (FEM) has been used extensively in simulation studies that could save test time, cost and workforce [19]. Numerical simulations have been widely used in the past for different AM processes to estimate temperature distributions, distortions, strains and residual stresses and are used as a fundamental tool for modeling different welding processes before actual experiments and for predicting and regulating material behavior [25–27]. The AM simulations are performed either using an implicit or explicit scheme. The implicit strategy takes an iterative approach until the required convergence conditions are fulfilled [28]. In contrast, the explicit method calculates a linear system behavior with a very small time-step. It is used for highly dynamic systems like crash or deformation processes without any iterative mechanism and convergence is often achieved by scaling time and mass [29–32].

The numerical simulation of WAAM is quite similar to the multi-pass welding. The transmission of heat and mass between the electric arc and substrate is regulated by the molten weld pool, with its complicated physical phenomena. The process is generally modeled with coupled thermo-mechanical welding analysis [33]. In WAAM numerical analysis, the heat transfer between the electric arc and weld pool is simulated through a heat-source model that defines heat generation in the welding region [34] for example a uniform Gaussian distributed heat source [35]. Whereas, material deposition is recognized using the element activation approach [36].

8.2.1   WAAM MODELING PROCEDURE

A brief introduction of the basic steps that are involved in the numerical modeling of the WAAM processes is given in this sub-section. This modeling technique can also be adapted for other AM processes depending on their process sequence and control parameters. Regardless of the mesh-based [37–39], or particles-based simulation methods [33,40,41] mostly WAAM simulations have been performed following material activation technique [42]. In the following text, the modeling procedure for a simple five-layered box has been presented (see Figure 8.5 (b)). The modeling and simulation analysis have been carried out using the software LS-Dyna, with Steel 309L as the substrate and build material.

8.2.2 MOVING HEAT SOURCE

The distribution of the energy in the weld pool during the WAAM analysis is governed by a moving heat source, i.e., Goldak's double ellipsoidal heat source [43]. The heat distribution in the forward and rearward regions of the weld pool can be found by the heat source equations (8.1) and (2) respectively [44]:

$$q_{forward}\left(a,b,c,t\right)=\frac{6\sqrt{3}f_fQ}{\pi\sqrt{\pi}wdc_f}e^{-3\left(\frac{a}{w}\right)^2}e^{-3\left(\frac{b}{d}\right)^2}e^{-3\left(\frac{c+v(\tau-t)}{c_f}\right)^2} \tag{8.1}$$

$$q_{rear}\left(a,b,c,t\right)=\frac{6\sqrt{3}f_rQ}{\pi\sqrt{\pi}wdc_r}e^{-3\left(\frac{a}{w}\right)^2}e^{-3\left(\frac{b}{d}\right)^2}e^{-3\left(\frac{c+v(\tau-t)}{c_r}\right)^2} \tag{8.2}$$

Where:

$q_{forward}$ and q_{rear} = weld source power density in forward and rearward of the heat source.
a,b,c = coordinates of the reference point P in the welding material
f_f = forward distribution function
f_r = rearward distribution function
c_f = forward direction weld pool
c_r = rearward direction weld pool

The rest of the parameters used to define Goldak's heat source are selected according to the previous work of Israr et al. [42] and Prasad et al. [45].

8.2.3 ELEMENT ACTIVATION

Element activation in WAAM processes follows two main techniques during the FE analysis. One of them is known as the quiet element technique [25] while the other is known as the inactive element technique [35]. In the quiet element technique, the whole model appears during the analysis whereas only the layers that receive the activation temperature are activated while the other elements remain in a quiet or ghost state [36]. However, in the inactive element approach, only the active elements appear in the model at a certain time based on their welding sequence. As the activation time and conditions reach, the elements started appearing and the model grows in size [46]. Both the techniques can be implemented in any AM process, depending on the model complexity, material deposition sequence and shape of the heat source

[47]. In the current work, the quiet element approach has been applied only to the build elements, which are activated according to their welding time.

8.2.4 BOUNDARY CONDITIONS

The numerical analysis of the WAAM process involves various boundary conditions that are incorporated in the model at the start of the simulation. Conduction with the substrate and surrounding layers, as well as convection and radiation to the air, influence heat flow during the simulation. For welding the layers, two types of contacts are used, i.e., tied contact and thermal contacts. The tied or mechanical contacts between the different layers and substrate are activated right from the start of the simulation while the time-dependent thermal contacts between the neighboring layers are activated according to the welding sequence. The activation of thermal contacts between layers is also associated with the termination of the heat loss through convection and radiation from the joined surfaces. The heat transfer between adjacent layers continues on contact activation but the thermal material properties of the new layer are activated only when the temperature under the heat source reaches above 1210 °C. Similarly, after the material has been thermally activated, the mechanical properties are activated when the temperature in the weld pool region rises over the melting point of the material, which in the case of steel is above 1400 °C. An illustration of boundary-condition activation in a WAAM model of a five-layered stacked block with four beads in each layer is depicted schematically in Figure 8.5.

In the given model, two types of contacts between the layers are established, i.e., tied or mechanical contacts and thermal contacts. The tied contacts between all the surfaces are activated from the start of the simulation and remain active until the end of the simulation. The other contacts are time-dependent thermal contacts that activate between the two joining surfaces only when the actual welding time of the layer

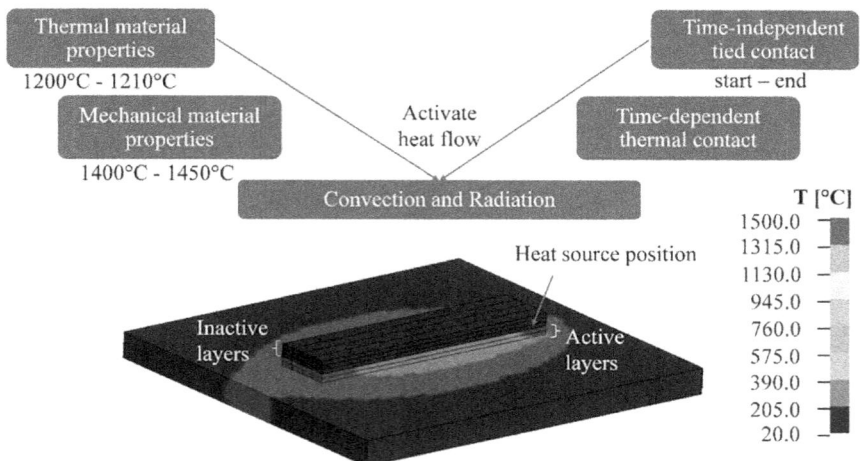

FIGURE 8.5 Demonstration of the Boundary Conditions during WAAM Simulations.

is achieved. Heat loss to the environment occurs exclusively through convection and radiation from the active layers. When the two joining surfaces make contact, the convection and radiation from the connected surface stop. The material activation boundary condition has been so configured that the thermal material properties are activated first as the energy from the heat source raises the temperature of the model over 1200 °C. To prevent convergence issues during the analysis, a temperature range of 1200 °C—1210 °C is specified for material activation. Similarly, as the model's temperature rises beyond its melting point, which for steel is between 1400 °C and 1450 °C, the material's physical properties are activated. The graphic below depicts the temperature distribution on the five-layered WAAM block and substrate. Active layers have already been printed, whereas inactive layers have elements that are still in the ghost or inactive state and have no influence on the model.

8.2.5 POTENTIAL OUTCOMES OF AM SIMULATIONS

Various outcomes can be achieved/observed through thermo-mechanical simulation analysis to improve the efficacy of the process before practical experimentation. For example, using numerical simulations, the strategies to optimize the substrate fixation can be established to reduce the residual stresses and minimize the distortion of the component [48]. The plastic deformation and residual stresses of the components can be evaluated in AM process through different path strategies [42]. Microstructural and phase change investigation of components can also be performed using thermo-mechanical analysis [49]. Furthermore, numerical simulations can be used to effectively optimize the welding path [50] and heat accumulation [36]. Using the numerical simulations, a thorough analysis of several processes and geometrical parameters influencing the analysis can also be assessed and regulated [51]. Hence, simulation approaches for AM processes have several advantages and can be used as a preliminary tool prior to real experimentation.

8.3 CONCLUSION

This study provides an overview of the different AM techniques for metals. The methods of powder bed fusion and wire-based DED processes have been discussed. The wire-based welding method, particularly WAAM, has been thoroughly studied, and its numerous advantages over other methods, as well as its process flow, have been illustrated. LS-Dyna software was employed to demonstrate the simulation modeling approach for a basic steel WAAM component. In the light of the prior work in the field of WAAM, prospective applications of numerical simulations have been presented. Following the above-mentioned facts on the AM and WAAM processes, a basic understanding and interest can be developed, which can be further strengthened by reading the author's extensive works and case studies.

8.4 ACKNOWLEDGMENTS

The authors would like to acknowledge funding through the project MALEDIF "Machinelles Lernen für die additve Fertigung", Antragsnnumer 85037495.

REFERENCES

[1] Bourell D, Kruth JP, Leu M, Levy G, Rosen D, Beese AM et al. Materials for additive manufacturing. CIRP Annals 2017;66(2):659–681.

[2] Mazumder J, Schifferer A, Choi J. Direct materials deposition: Designed macro and microstructure. *Materials Research Innovations* 2016;3(3):118–131.

[3] Sachs ME, Haggerty JS, Cima MJ, Williams PA. *Three dimensional printing techniques* (US Patent; 5204055A), 1993.

[4] Zhang Y, He X, Han J, Du S. Ceramic green tape extrusion for laminated object manufacturing. *Materials Letters* 1999;40(6):275–279.

[5] Melchels FPW, Feijen J, Grijpma DW. A review on stereolithography and its applications in biomedical engineering. *Biomaterials* 2010;31(24):6121–6130.

[6] Zou Y, Li C-H, Liu J-A, Wu J-M, Hu L, Gui R-F et al. Towards fabrication of high-performance Al2O3 ceramics by indirect selective laser sintering based on particle packing optimization. *Ceramics International* 2019;45(10):12654–12662.

[7] Guo N, Leu MC. Additive manufacturing: Technology, applications and research needs. *Frontiers in Mechanical Engineering* 2013;8(3):215–243.

[8] Lim S, Buswell RA, Le TT, Austin SA, Gibb AGF, Thorpe T. Developments in construction-scale additive manufacturing processes. *Automation in Construction* 2012;21:262–268.

[9] Baufeld B, van der Biest O, Gault R. Additive manufacturing of Ti—6Al—4V components by shaped metal deposition: Microstructure and mechanical properties. *Materials & Design* 2010;31:S106–S111.

[10] Brandl E, Baufeld B, Leyens C, Gault R. Additive manufactured Ti-6Al-4V using welding wire: Comparison of laser and arc beam deposition and evaluation with respect to aerospace material specifications. *Physics Procedia* 2010;5:595–606.

[11] Jiang J, Xu X, Stringer J. Optimization of process planning for reducing material waste in extrusion based additive manufacturing. *Robotics and Computer-Integrated Manufacturing* 2019;59:317–325.

[12] Delgado Camacho D, Clayton P, O'Brien WJ, Seepersad C, Juenger M, Ferron R et al. Applications of additive manufacturing in the construction industry—A forward-looking review. *Automation in Construction* 2018;89:110–119.

[13] Ding D, Pan Z, Cuiuri D, Li H. A multi-bead overlapping model for robotic wire and arc additive manufacturing (WAAM). *Robotics and Computer-Integrated Manufacturing* 2015;31:101–110.

[14] Dutta B, Froes FH. The Additive Manufacturing (AM) of titanium alloys. *Metal Powder Report* 2017;72(2):96–106.

[15] Kruth J-P, Levy G, Klocke F, Childs THC. Consolidation phenomena in laser and powder-bed based layered manufacturing. *CIRP Annals* 2007;56(2):730–759.

[16] Sreenathbabu A, Karunakaran KP, Amarnath C. Statistical process design for hybrid adaptive layer manufacturing. *Rapid Prototyping Journal* 2005;11(4):235–248.

[17] IMAM. *Inside metal additive manufacturing* [September 28, 2021]; Available from: www.insidemetaladditivemanufacturing.com/blog/wire-feed-additive-manufacturing.

[18] Williams SW, Martina F, Addison AC, Ding J, Pardal G, Colegrove P. Wire + Arc additive manufacturing. *Materials Science and Technology* 2015;32(7):641–647.

[19] Gornyakov V, Sun Y, Ding J, Williams S. Computationally efficient models of high pressure rolling for wire arc additively manufactured components. *Applied Sciences* 2021;11(1):402.

[20] Cheepu M, Lee CI, Cho SM. Microstructural characteristics of wire arc additive manufacturing with inconel 625 by super-TIG welding. *Transactions of the Indian Institute of Metals* 2020;32(2019):641.

[21] Ding D, Pan Z, Cuiuri D, Li H. Wire-feed additive manufacturing of metal components: Technologies, developments and future interests. *The International Journal of
Advanced Manufacturing Technology* 2015;81(1–4):465–481.

[22] Bambach M, Sizova I, Emdadi A. Development of a processing route for Ti-6Al-4V
forgings based on preforms made by selective laser melting. *Journal of Manufacturing
Processes* 2019;37:150–158.

[23] Addison A, et al. Manufacture of complex titanium parts using wire + Arc additive
manufacture. *International Titanium Association*, Cranfield University. 2015.

[24] Bambach M, Sizova I (eds.). Hot working behavior of selective laser melted and laser
metal deposited Inconel 718. *AIP Conference Proceedings* 1960, 170001 (2018).

[25] Chiumenti M, Cervera M, Salmi A, Agelet de Saracibar C, Dialami N, Matsui K.
Finite element modeling of multi-pass welding and shaped metal deposition processes.
Computer Methods in Applied Mechanics and Engineering 2010;199(37–40):2343–2359.

[26] Rodrigues TA, Duarte V, Miranda RM, Santos TG, Oliveira JP. Current status and perspectives on wire and arc additive manufacturing (WAAM). *Materials (Basel)* 2019;12(7).

[27] Gu J, Wang X, Bai J, Ding J, Williams S, Zhai Y et al. Deformation microstructures
and strengthening mechanisms for the wire+arc additively manufactured Al-Mg4.5Mn
alloy with inter-layer rolling. *Materials Science and Engineering: A* 2018;712:292–301.

[28] Harewood FJ, McHugh PE. Comparison of the implicit and explicit finite element methods using crystal plasticity. *Computational Materials Science* 2007;39(2):481–494.

[29] Hu X, Wagoner RH, Daehn GS, Ghosh S. Comparison of explicit and implicit finite
element methods in the quasistatic simulation of uniaxial tension. *Communications in
Numerical Methods in Engineering* 1994;10(12):993–1003.

[30] Choi H-H, Hwang S-M, Kang YH, Kim J, Kang BS. Comparison of implicit and explicit
finite-element methods for the hydroforming process of an automobile lower arm. *The
International Journal of Advanced Manufacturing Technology* 2002;20(6):407–413.

[31] Sun JS, Lee KH, Lee HP. Comparison of implicit and explicit finite element methods for dynamic problems. *Journal of Materials Processing Technology* 2000;
105(1–2):110–118.

[32] Yang DY, Jung DW, Song IS, Yoo DJ, Lee JH. Comparative investigation into implicit,
explicit, and iterative implicit/explicit schemes for the simulation of sheet-metal forming processes. *Journal of Materials Processing Technology* 1995;50(1–4):39–53.

[33] B. K. Mishra, Raj K. Rajamani. The discrete element method for the simulation of ball
mills. *Applied Mathematical Modelling* 1992;16.

[34] Montevecchi F, Venturini G, Scippa A, Campatelli G. Finite element modelling of
wire-arc-additive-manufacturing Process. *Procedia CIRP* 2016;55:109–114.

[35] Li T, Zhang L, Chang C, Wei L. A uniform-Gaussian distributed heat source model
for analysis of residual stress field of S355 steel T welding. *Advances in Engineering
Software* 2018;126:1–8.

[36] Israr R, Buhl J, Bambach M. A study on power-controlled wire-arc additive manufacturing using a data-driven surrogate model. *The International Journal of Advanced
Manufacturing Technology* 2021;117(7).

[37] Donea J, Huerta A, Ponthot J-P, Rodríguez-Ferran A. Arbitrary Lagrangian—Eulerian
methods: Chapter 14. Universitéde Li'ege, Li'ege, Belgium, Universitat Polit'ecnica de
Catalunya, Barcelona, Spain. E. Stein, R. de Borst and T.J.R. Hughes, Wiley & Sons; 2004.

[38] Zhang L, Michaleris P. Investigation of Lagrangian and Eulerian finite element methods for modeling the laser forming process. *Finite Elements in Analysis and Design*
2004;40(4):383–405.

[39] Kucharik M, Liska R, Vachal P, Shashkov M. Arbitrary Lagrangian-Eulerian (ALE)
methods in compressible fluid dynamics. *Programs and Algorithms of Numerical
Mathematics* 2006;13:178–183.

[40] Liang D, Jian W, Shao S, Chen R, Yang K. Incompressible SPH simulation of solitary wave interaction with movable seawalls. *Journal of Fluids and Structures* 2017;69:72–88.

[41] Belytschko T, Lu YY, Gu L. Element-free Galerkin methods. *International Journal for Numerical Methods in Engineering* 1994;37(2):229–256.

[42] Israr R, Buhl J, Elze L, Bambach M (ed.). *Simulation of different path strategies for wire-arc additive manufacturing with Lagrangian finite element methods.* Cottbus: Israr, Rameez; 2018.

[43] Goldak J, Bibby M, Moore J, House R, Patel B. Computer modeling of heat flow in welds. *Metallurgical Transactions B* 1986;17(3):587–600.

[44] Goldak J, Chakravarti A, Bibby M. A new finite element model for welding heat sources. *MTB* 1984;15(2):299–305.

[45] Bhusan Prasad B. *Simulation of heat source in gas metal arc welding by using.* LS-PrePost (Tutorial); 2018:1–32.

[46] Lindgren L-E, Runnemalm H, Nsstrm MO. Simulation of multipass welding of a thick plate. *International Journal for Numerical Methods in Engineering.* 1999;44(9):1301–16.

[47] Dantin M, Furr W, Priddy M (eds.). Towards an open-source, preprocessing framework for simulating material deposition for a directed energy deposition process. In *Proceedings of 29th International Solid Freeform Fabrication Symposium*; Austin, Texas, USA: University of Texas; 2018.

[48] Israr R, Buhl J, Bambach M. numerical analysis of different fixation strategies in direct energy deposition processes. *Procedia Manufacturing* 2020;47:1184–1189.

[49] Buhl J, Klöppel T, Merten M, Haufe A, Rameez I, Bambach M. *Numerical prediction of process-dependent properties of high-performance Ti6Al4 in LS-DYNA.* ESAFORM; 2021.

[50] Fügenschuh A, Bambach M, Buhl J (eds.). Trajectory optimization for wire-arc additive manufacturing: In *Operations Research Proceedings 2018*. Cham: Springer International Publishing; 2019.

[51] Buhl J, Israr R, Bambach M. Modeling and convergence analysis of directed energy deposition simulations with hybrid implicit/explicit and implicit solutions. *Journal of Machine Engineering* 2019;19(3):95–108.

9 Additive Manufacturing Based Rapid Tooling

Syed Waqar Ahmed, Khurram Altaf,
G. Hussain, Junaid Qayyum and Adeel Tariq

CONTENTS

9.1 INTRODUCTION

The present-day requirement of industries includes swift and economic manufacturing of complex and intricate components. Metal injection molding (MIM) is among the popular processes for swift production of intricate components [1]. Conventionally, molds for MIM process are made through machining, and owing to the high manufacturing cost, the use of machined molds is restricted to mass production. The use of machined metal molds could be redundant for certain applications demanding low volumes of MIM parts, especially prototyping, design validation, analysis and other upstream processes of permanent mold making. Additionally, the customized and complex geometric features require intensive machining, thereby making the overall process time-consuming and exorbitant.

Fused deposition modeling (FDM) has an inherent advantage to manufacture complex molds accurately, while being virtually independent of the geometrical features and constraints. Therefore, FDM can potentially be used to produce MIM molds for part production. Although, the FDM-made molds cannot be used for

DOI: 10.1201/9781003220985-9

very high number of MIM parts, however, for low volume demands of MIM parts, FDM-made polymer molds could be economical and promising. In applications where design change is frequent and MIM parts are required in low volume, the FDM polymer molds could be potentially suitable. Additionally, in high-volume production demands, rigorous analysis of design is inevitable before finalization for permanent mold making. If the design analysis is overlooked and permanent metal mold is machined, any modification in the design at this stage, could practically be equivalent to machining of a completely new mold. Therefore, in the phase of design validation and form and fit analysis the design may undergo several modifications before finalization, to avoid any further complications in permanent mold [2]. Polymer molds having threshold values for strength, roughness and accuracy as 500 MPa, 1µm and 0.1 mm respectively are believed to be favorable for MIM [3].

Rapid tooling (RT) is defined as development of mold through additive manufacturing (AM)/3D Printing (3DP), typically for injection molding (IM) process [4]. It is an off-shoot development of rapid prototyping (RP) that was developed in 1980s [5]. RT process can either involve AM processes to make mold, called direct RT [6] or can use RP components as a pattern for mold development, called indirect RT [7]. RT has reduced time to market and RP has reduced development time [8]. RP is preferred for development of only a single part whereas RT is a preferred choice for production of a small volume of parts [9]. RP can be useful for up to five components, whereas RT is preferred where more parts are required [10]. RP is more convenient for direct manufacturing of end-use applications as a customized part [11].

In molding industries, RP is relatively popular in manufacturing of sacrificial patterns for casting or silicone molding. However, for making molds for MIM, RT is superior to RP. Both RT and RP follow the similar generic process of AM, which starts by slicing of a 3D model of mold into thin layers followed by gradual deposition of each slice over the previous layer in a sequential manner [12]. Contrary to conventional machining, where a large block is machined to remove material, this process selectively deposits material according to sliced computer aided design (CAD) model. The process is quick, economical and accurate for intricate geometries [13]. In a comparative analysis, it was found that due to RT, time and cost of mold manufacturing have been reduced by 90% and 70% respectively [14]. Since cost and time in additive manufacturing are virtually independent of intricacy of features, nevertheless, the machining cost and time could be challenging due to intricacies and complexities of the features.

An inherent downside of FDM-made molds is that the surface quality is not very high. This could demand post-processing to improve the surface of molded part, thus making overall MIM process time-consuming. A precise control and careful combination of FDM processing parameters is a quick and easy solution to considerably control the surface quality [15]. For advance and superior surface control, composite filaments are promising for sustainable mold development. The addition of foreign particles or additives in the FDM polymers could improve mechanical strength and surface quality as well [16]. Another limitation of FDM-made molds is the presence of anisotropy, typically in mechanical properties. The strength and surface properties are highly direction dependent and thus the properties could not be generalized for the mold [17]. The anisotropy could be mainly controlled by parametric optimization [18] and composition enhancement [19]. However, the effects of composition enhancement pertaining to anisotropy are more prominent and effective. Moreover, metal additives are promising for strength enhancement whereas polymer additives augment isotropy and surface of the resulting composites.

FDM molds are polymer in nature and the MIM feedstock contains considerable amount of polymer as binder [20]. As MIM operation is performed at high temperature and pressure, there occurs likelihood of adhesion between the feedstock polymer and the mold cavity wall. This adhesion could result in part failure during release [21]. To address this limitation, metal coating on FDM-made polymer molds could be a promising remedy [22]. Metal plating on polymer mold could serve as a barrier to prevent interfacial adhesion between the feedstock and mold. Additionally, it can protect mold surface from rapid deterioration as well. Electroless metal coating has been studied by many researchers, especially on common RT polymers like acrylonitrile butadiene styrene (ABS) and poly lactic acid (PLA) [23]. Therefore, a mold can be coated with metal before subjecting to MIM for improved performance [24]. Some studies reported improvement of part release in metal plated FDM molds by enhancing the design of mold insert [25]. Enhancements in other parameters, typically composition of mold, are also fruitful in improvement of performance of FDM-made polymer molds for metal injection molding applications [26].

This chapter reviews surface enhancement, build material, and design as an effective and efficient post-processing approach to successfully augment rapid tooling and MIM technology, which is a major grey area in some of the available literature. Additionally, the chapter also motivates the prospective readers to investigate more efficient techniques to plate polymer molds with metals and to optimize the overall MIM process.

9.2   FUSED DEPOSITION MODELING FOR RAPID TOOLING

Fused deposition modeling (FDM) is a 3D Printing/Additive Manufacturing process of depositing material layer-by-layer through heated nozzle, whose position is controlled typically in two axes. The deposition in third axis is carried out by descending of print platform [27]. The FDM machine takes the file in .stl format and horizontally slices it to sections for printing. Presently many non-toxic and odorless materials are available for FDM, which have better environmental effect and make the process convenient for use in closed environment as well [28]. The technology was introduced by Scott Crump in 1980s and first commercialized in 1990s by a company founded by himself [29]. It is the most popular AM technology after stereolithography (SLA) [30] however, still it can be classified as one of most popular processes for polymer RT [9].

As already mentioned, mold making through FDM is mainly aimed as a quick and economical approach for design validation, prototyping, low-volume demands, and form and fit analyses as upstream processes before commitment for permanent hard tooling. The use of FDM-made molds is becoming popular and common in casting, wax and plastic molding industries [31]. Nevertheless, in MIM industries,

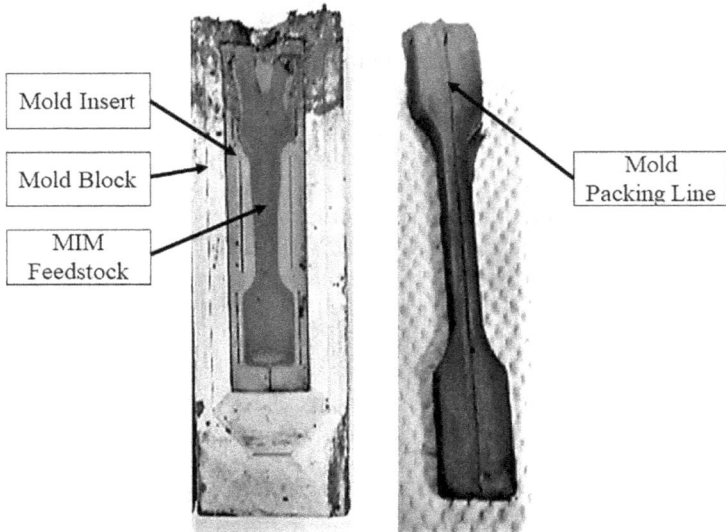

FIGURE 9.1   Use of FDM Molds of ABS Material for Copper Injection Molding [35].

there exists debate regarding reliability and sustainability of direct use of FDM-made polymer molds [22]. Yet enhancement techniques are promising to make FDM polymer molds suitable for their direct use in MIM.

Customized mold making through reverse engineering of the image is popular in biomedical prostheses applications. The biomedical images are converted to CAD files through reverse engineering and then processed for customized mold making. This scheme has considerable reduced time, cost and skills for manufacturing of customized implants [32]. Use of FDM-made molds for successful manufacturing of patient-specific biomedical implants has been reported in some studies, which are source of motivation for the prospective researchers to further enhance the overall manufacturing process [33].

The selection of mold material and manufacturing method is mainly governed by cost, manufacturability, and tooling requirements. The manufacturing process, be it FDM or otherwise is selected once these parameters are optimized and justified accordingly. Some studies presented logics to compare and evaluate the selection criterion of FDM process among other mold manufacturing processes for making of MIM molds [34].

9.3   PROCESS PARAMETER'S ENHANCEMENT FOR RAPID TOOLING

Surface roughness is an inherent downside of layered manufacturing processes and FDM being no exception [36]. This could hamper filling of mold cavity and surface quality of MIM part. Careful selection of FDM deposition parameters is an economical and prompt choice to control surface degradation. The underlying advantage of parametric enhancement over the post-processing is that the parametric enhancement does not require any additional skill or material to be deposited or infused, as in the case of material enhancement and surface coating. A good parametric combination implies minimum number of start-stops and turns in the part. The parametric enhancement has been reported to improve surface quality, mechanical strength and manufacturing cost of the mold [37]. The surface roughness occurs in the FDM-made molds due to the reason that the material is deposited in discreet layers. There always lies the probability of improper fusion between the raster layers, which results in the surface roughness [38]. The staircase effect could be estimated as product of raster thickness and tangent of the vertical inclination CAD model [39]. The thinner raster layer could contribute to improved surface finish on one hand but could increase manufacturing time on the other hand as well. Therefore, a compromise needs to be established for optimization of the overall process [40]. Therefore, the interior layers of the part could be deposited with thick layer to save build time while the exterior could be deposited with fine and thin layers [41].

Enhancement of the tool-path parameters has been reported as helpful to improve the surface quality of FDM-made RT, due to which an inherent challenge in layer-by-layer deposition could be addressed. The continuous spiral deposition helps to avoid gaps and to improve the surface quality of tools. This enhancement has successfully been used in a case study to manufacture molds for highly customized tracheal stent [42].

For a critical analysis, the expression of process parameters as function of the product characteristics is important. Common FDM parameters are raster width, layer height and deposition speed. In a case study of parametric analysis, the layer thickness was reported as relatively more prominent than raster width and speed, in controlling

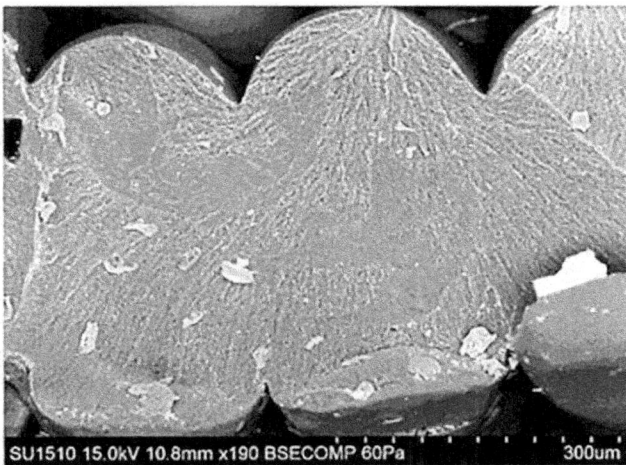

SU1510 15.0kV 10.8mm x190 BSECOMP 60Pa   300um

FIGURE 9.2 Circular FDM Filament Becoming Ellipsoid Due to Prolong Cooling Time [44].

the surface quality [43]. Nevertheless, it may be noticed that these correlations are magnitude dependent. For the different selection groups of process parameter values, the layer thickness may not be the most significant process parameter to control surface quality. However, the parametric analysis could still estimate the general trend in the product character as a function of process parameters. Process parameters like raster speed and part geometry also play a role in surface distortion and geometrical inaccuracies. If the deposition speed is quick and a layer is deposited over the other before the bottom layer has solidified properly, the incoming layer could increase the solidification time. This would make the round raster as elliptical and lead to curling or volumetric errors. Bottom layers are more prone to volumetric errors, compared to the top layers, nevertheless, mechanical strength of resulting part would be comparatively superior due to proper fusion of the layers [44]. Higher raster width and lower layer thickness could minimize the deviation in thickness, whereas deviation in length and width could be controlled by high layer thickness and medium raster width [45].

Regarding the effects of parameters like air gap and model temperature, some the studies reported that their effects are less prominent on the surface quality of the parts because their effects are dictated by the raster width, layer thickness and raster speed to some extent. Yet very less or overlapping raster layers and rapid cooling would result in degraded surface quality [45]. Build orientation and support material quantity are likely to effect the surface quality and hardness, typically for overhang and down-facing geometries [46].

A small scale post-processing could sizably contribute to the improvement in surface quality of FDM parts. In a comparative study, investigating surface quality as function of orientation angle, part density, vapor deposition time, and cooling time. It implies that the effect of post-processing may not be very significant in augmenting mechanical strength, nevertheless, the effects on surface quality is dominated by post-processing compared to deposition parameters. Smoothing time has highest effect on the surface quality among orientation angle, part density, smoothing time, vapor deposition time, and cooling time [33]. For some particular FDM parts, orientation angle may not always be significant factor in achieving surface quality. Part density and cooling time have relatively prominent influence on the surface hardness, which is an added quality of the molds [47].

The surface roughness could be function of many parameters; however, it could be expressed as function of pre-cooling time and post-cooling time as;

$$R_a = \frac{T_i}{T_f^2} \cdot K$$

K denotes the constant of proportionality and T_i and T_f denote pre-cooling time and post cooling time respectively. Assuming T_s as smoothing time, value of K could be further approximated through the following expression [48];

$$K = -0.075T_s^2 + 4.16T_s + 37.63$$

Machining could also be a promising post-processing technique to mitigate staircasing and to improve the surface quality of FDM molds [49]. Besides conventional machining, use of hot cutter could generate superior surface because it fuses the surface above glass transition temperature and smoothens the raster layers simultaneously. The hot cutter machining could make surface finish of the order of 0.3 µm [50]. Another advantage of hybridizing hot cutter machining (HCM) with FDM is that as the layer is deposited in FDM, the HCM tool immediately smoothens the desired regions before they become inaccessible due to intricacies. Likewise, integration of two FDM units to make a single manufacturing system provides the capability to use multi-material configuration for the fabrication of molds. Rather than using enhanced material for full mold manufacturing, only the top surface of the mold could be deposited with the enhanced material to improve the surface quality. Apparently, it could be a concern that multi-materials would not be joined properly, however, the experiments show that the deposition occurs at elevated temperature, due to which the joining is adequate. The mechanical and surface properties of parts made through multi-technology FDM units are comparable to the conventional single material manufacturing units [51,52].

9.3.1 SUMMARY

- If the layer width were small, the number of layers would be increased to fill the gap. This means that the part would remain heated for comparatively longer time. Moreover, an increase in heating and cooling cycle would be observed. This would help to achieve proper joining of layers and better surface. However, too small layer thickness would shrink the part due to cooling and over joining, which would introduce residual stresses and decrease mechanical strength. Moreover, the surface could also be degraded due to overlapping or over joining of the layers. Therefore, moderately low layer width is recommended for a superior surface finish.
- Small raster angles may contribute to mechanical properties lengthwise; however, residual stress may be introduced in the parts if the raster length is too much. Large raster angles may increase the shrinkage and decrease mechanical strength in transverse direction but improve the surface properties. Therefore, larger raster angles are suitable for proper surface finish.
- Too less air gap may improve diffusion and mechanical strength between the layers but would reduce surface quality due to overlap. Too high air gap may result in improper joining and crests on the surface. Therefore, moderately low air gap is recommended for a superior surface and mechanical properties.
- Comparatively thick rasters contribute to the heat accumulation at the joining surfaces, which helps in proper bonding, mechanical strength and surface quality [53].

9.4 COMPOSITION ENHANCEMENT FOR RAPID TOOLING

Presently, the leading materials in FDM machines are acrylonitrile butadiene styrene (ABS) and nylon [54]. Addition of foreign particles in the base polymers could be responsible for improvement in overall performance of composite material [55].

Common additives are metals, ceramics and polymers, which could be introduced in the base material to develop graded materials to achieve customize performance of the molds. Another objective of material enhancement is to control the inherent challenges of FDM process, typically anisotropy, mechanical properties and surface quality. If W_{add} denotes the weight grams of additive, W_{bas} denotes weight grams of base material, W_p denotes weight percent of plasticizer and W_s denotes weight percent of surfactant, the weight of composite W_c can be computed through following relation [26].

$$W_c = \frac{W_{add} + W_{base}}{1 - \left(W_p\% + W_s\%\right)}$$

If σ and V denote yield strength and volume fraction respectively and the subscripts m, r and c are for matrix, reinforcing agent and composite respectively, the yield strength of composite can be given by the following expression [56].

$$\sigma_c = \sigma_m V_m + \sigma_r V_r$$

If T and ω denote glass transition temperature and weight fraction respectively, and c, 1 and 2 are subscripts for composite, first and second additive respectively, the glass transition temperature of resulting blend is given by the following equation [57].

$$\frac{1}{T_c} = \frac{\omega_1}{T_1} + \frac{\omega_2}{T_2}$$

For the customized and graded applications, the composition could be tailored according to the law of averages [58]. For instance, the density of blend is estimated in terms of weight fraction as;

$$\rho_c = x_1 \rho_1 + x_2 \rho_2$$

From the above-mentioned mathematical expressions, it could be observed that the mechanical properties could be augmented through the addition of foreign particles in the base material. Similarly, enhancement in densification and glass transition temperature could retain surface quality at comparatively higher temperature and pressure conditions, thereby rendering the mold feasible for use in relatively hostile and non-conventional MIM processes with novel or tailored feedstock.

9.4.1 Polymer-Ceramic Composites for Rapid Tooling

Usually, metal oxide particles size range is ~ 50 μm. Addition of such large particles may not dramatically alter the mechanical properties of the composite. Nevertheless, addition of nanoparticles makes the composite brittle because the agglomerated particles decrease elongation at fracture due to cohort of stress-concentrators. Therefore, the effect of added particles as stress-concentrators dominates and decreases ultimate tensile strength (UTS) of the composite. The functionalization of ceramic additives, typically TiO_2 improves the uniformity of distribution of particles in the matrix and therefore mechanical property anisotropy could be controlled [59].

Contrary to polymer-polymer composites where generation of micro-voids is prominent, the ceramic-polymer composites have comparatively less micro-voids because the ceramic particles are of uniform size. The fracture in ZXY printed specimens is similar to fracture of pure ABS printed specimens, which implies that addition of ceramic particles did not significantly affect the raster adhesion or mechanical property anisotropy. Rather the ceramic particles control mechanical property anisotropy of the resulting composite. Additionally, presence of these particles contributed to control micro-voids and to arrest the cracks.

Addition of ZnO nano-rods rendered the fracture mode to brittle due to presence of micro-voids. Presence of intermingling was observed in the SEM inspection; nevertheless, the tensile properties were dictated by micro-voids in the composite. The ZnO blend did not enhance tensile properties or mechanical property anisotropy because non-spherical geometry of the additive resulted in generation of micro-voids in the composite. This implies that the uniformity in the geometry of additive is equally significant to achieve superior isotropy and strength [60].

Addition of $SrTiO_3$ in ABS mitigated tensile and mechanical property anisotropy of the resulting blend. The SEM reveals presence of micro-voids in the printed specimen, which resulted in brittle failure.

Compared to pure ABS, the composite of ABS-Alumina blends resulted in increase in intermingling of rater layers was obvious in the ZXY printed specimens, nevertheless the presence of micro voids and the behavior of alumina particles as crack initiators, synergistically dominated the intermingling effect. Characteristic brittle fracture was observed due to presence of micro voids in the composite [59].

9.4.2 POLYMER-METAL COMPOSITES FOR RAPID TOOLING

Three main factors, which determine the properties of polymer-metal composite, are the particle size, geometry and percentage. Addition of fine particles is responsible for increased fracture elongation and increased toughness. However, the ultimate strength and yield strength depreciates. Coarse and larger particles are responsible for brittleness and high tensile strength. This behavior is exhibited due to the reason that large particles result in large fracture energies and increase in crack precursors. Smaller particles exhibit superior packing due to which toughness is increased. The increase in amount of metal particles results in decrease in toughness, brittleness, elongation, and tensile strength. This behavior of the composite is evident from the illustration provided in Figure 9.3 This is attributed to ineffective interaction when the particles are placed close to each other. Each added particle bears void, which is unavoidable during making of composites therefore, each particle could be sought as the crack precursor. Moreover, the bond between metal particles in composites is not that strong as the polymer-polymer bond. Therefore, the decrease in the tensile properties is noticeable [26].

The mold manufactured through FDM process using iron-nylon filament, was successfully used for 70 injection molding cycles using LDPE and ABS feedstock [26].

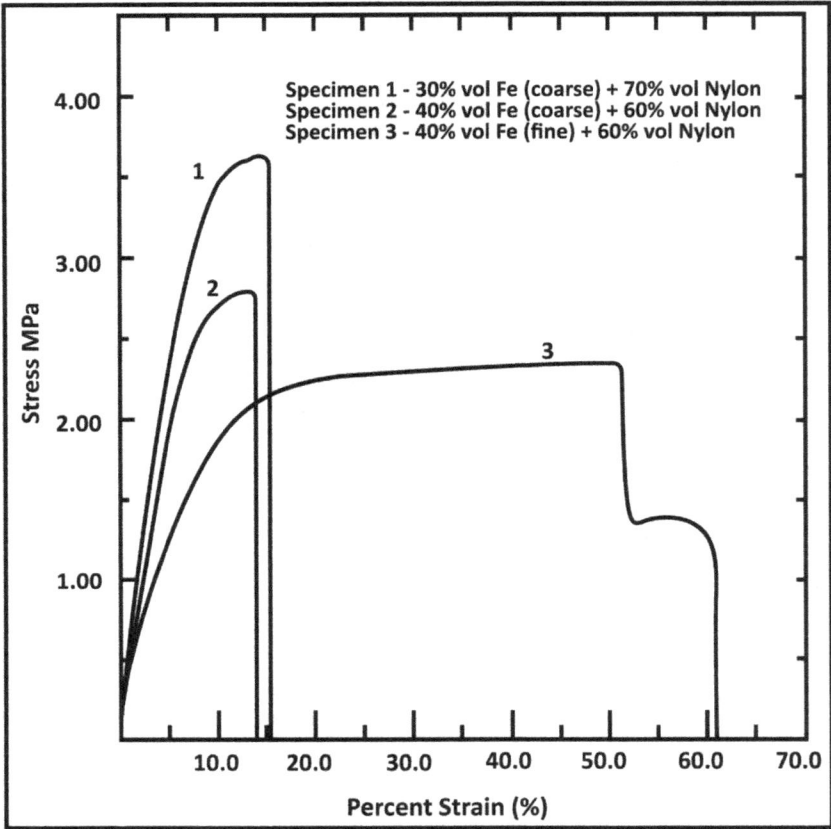

FIGURE 9.3   The Effect on Tensile Properties of Iron-Nylon Composite Due to Addition of Iron Particles [26].

The overall performance of this composite mold was comparable to pure nylon mold. Additionally, the composite filament was also used to manufacture intricate inserts and functional parts for related applications.

Other introductory studies reported the copper-ABS composite fabrication and observed increase in storage and loss modulus, proportional to amount of added copper [61]. The modeling of deposition behavior of the copper-ABS composite is a herald that it is suitable for FDM mold making [62]. Successful use of copper-ABS composite and iron-ABS composites provides confidence in reliability of using them in molding industries [63].

9.4.3   POLYMER-POLYMER COMPOSITES IN RAPID TOOLING

Addition of jute fibers in ABS leads to generation of voids in the composite due to burning of jute while filament processing. These voids are responsible for decrease

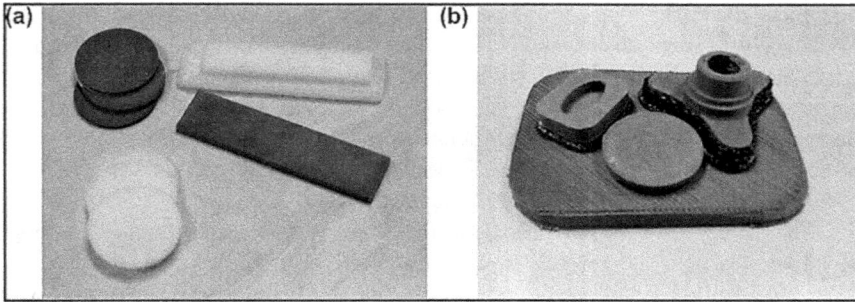

FIGURE 9.4   (a) Test Samples (b) Prototype Parts Made through Iron-ABS Composite Filament [63].

of effective cross-sectional area and increase of stress-concentrators. The voids rendered the fracture as brittle because percentage elongation was decreased to one-fifth. Additionally, UTS of composite was decreased to 8.63 MPa compared to 17.73 MPa for ABS filament. However, the failure analysis shows that interfacial adhesion is superior to the strength of resulting composite material.

When 2% of MayaCorm Blue composite was added to ABS, the resulting composite manifested similar failure compared to ABS/Jute composite. The specimens printed in XYZ direction manifested maximum percentage elongation of 8.86 % compared to 8.6% of pure ABS. nevertheless, the fracture in ZXY printed specimens was brittle due to generation of voids in the composite, which were attributed to decomposition of indigo component of composite during filament processing. The raster adhesion was superior, and failure was observed within the raster layer [64].

Most of metal or ceramic additives in ABS polymer render the resulting composted blend brittle. However, biopolymer blend composed of ABS and SEBS in 80:20 weight ratio exhibits comparatively ductile fracture. Compared to ceramic-polymer blends, micro voids and mechanical property anisotropy were controlled in the polymer-polymer blend. The smooth surface in ZXY printed specimens indicates rapid crack propagation [65].

5 wt. % addition of SEBS polymer in ABS has contributed to decrease in mechanical anisotropy of the resulting binary polymer composite [66]. Binary polymer of ABS and SEBS in 50:50 and 80:20 also contribute to mitigation in surface roughness at selected raster angles. Additionally, ternary polymer composite composed of ABS, UHMWPE and SEBS in ratio of 75:25:10 contributes to superior surface finish of inclined surfaces. This decrease contributes significantly to mitigate an inherent challenge of FDM machines. From SEM analysis, it was observed that intermingling of raster layers contributes to the decrease in surfer roughness [67].

It may however, be noticed from careful comparison of the tensile testing results that the UTS of blend is less compared to pure ABS, nevertheless the inherent limitations of the FDM process like surface smoothness and mechanical property anisotropy are considerably mitigated [61].

These materials are however, limited in strength and their use is restricted for a few IM cycles. Considering constraint of low deflection temperature of machine, researchers have developed polymer based filaments, which carry metal powders embedded/infused in the filament [26,68]. The polymer composite flows from the nozzle, carrying metal particles within itself. The polymer matrix can be removed after sintering if required. Composite polymer RT can potentially be superior to conventional filaments due to a significant metal content, typically 80% by weight. Elastic modulus increases with increase in metal contents in PLA, however converse is true for tensile strength and fracture toughness [69]. Composite RTs can survive for 1,000 IM cycles with polypropylene feedstock [70]. Presently many composite resins like polymers and epoxies are being used for RT [71].

Polyether Imide (trade name as Ultem 9085) is a high-performance polymer which can be used to make end-use components. Its normal extrusion and preset tip temperatures are 330–360°C and 380°C respectively. The shear rate has been calculated as 200/s, at 0.4mm diameter of extruder tip. Its density is 1.27 g/cm^3 with molecular formula as $C_{37H24O6N2}$, it has been used in FDM process effectively and efficiently [72]. Some of the researchers have also reported the use of metal and ceramic infused filaments. Filaments containing hydroxyapatite, alumina, stainless steel and piezoelectric materials have also been used in manufacturing of direct end-use components from FDM [73]. Additives in FDM filaments can lead to brittleness, however further addition of linear polymers can render them ductile [74]. Thermo-tropic liquid crystalline polymers (TLCP) are added for high aspect ratio filaments so as to keep them from fracture during extrusion [75]. Introduction of styrene ethylene butadiene styrene decreased anisotropy from 47% to 22% [65]. In a self-generated functionally graded filament comprising of nylon-6, Al and Al_2O_3 in ratios of 60:30:10, it was found that presence of Al_2O_3 could significantly improve micro-hardness and wear resistance [76]. Sustainability of iron infused nylon (mixed in 30:70 and 40:60) was investigated for RT application. Maximum of 70 parts of polyethylene were successfully ejected form thus made rapid tool [26]. Polypropylene binder was used to infuse alumina, silica and titanium oxide to manufacture components or tools followed by infiltration and de-binding with aluminum to evaluate the surface characteristics. For the optimal combination of ingredients to obtain acceptable roughness, parametric investigation can be done using DOE [77]. Accuracy and reproducibility capabilities of using stainless steel 17–4 PH powder were studied to develop composite filament for FDM based RT and direct end-use parts. Different filaments, each with 40%, 30%, 20% and 10% concentration of 58% mixture of 17–4PH SS were prepared. A lug fit and a wrench were made using composite filament and accuracy and reproducibility were compared against RTV molding. Performance of composite filament was satisfactory in the selected case study. The approach however, can be extended on mechanical properties, lead time and cost analyses of parts [78].

Results from dynamic modeling [79] developed already for single composition filaments are not applicable for composite filaments, therefore improved investigations were conducted for accurate dynamic modeling of flow behavior of composite filament [80]. Flow viscosity in tailored FDM composite filament was studied, in which amount of copper was increased from 10% to 40%, in increments of 10% and ABS decreased from 85% to 55% in decrements of 2.5%. Best flow behavior was obtained when composite ingredients contained ABS, copper, binder and surfactant in 62.5:30:6:1.5 ratio [81]. Study on evaluation of dynamic mechanical properties of copper filled ABS was also reported. Trends of complex viscosity, loss factor, storage modulus and loss modulus were studied. It was learnt that loss factor increased up to glass transition temperature and decreased afterwards. Rest of the parameters decreased with increase in temperature due to increasing vibration among the molecules. Composite filaments posses superior properties and therefore can potentially be significant in direct RT using FDM [26,63,67]. In another study by Nikzad et al. [66] iron powder (average size 45 μm) was loaded in ABS in 10:90 to generate composite feedstock for FEA analysis of its potential use in direct RT. Glass transition temperature and thermal conductivity were determined analytically and storage

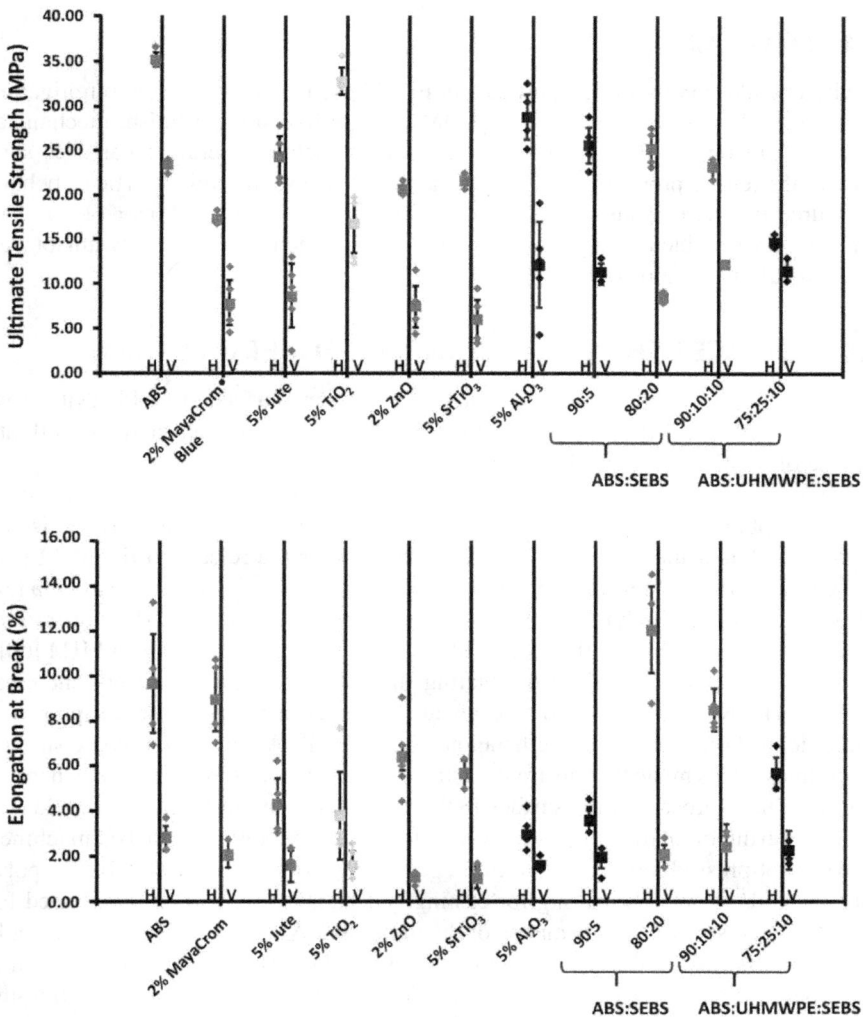

FIGURE 9.5 Effects of Additives On (a) UTS and (b) Elongation [65].

modulus was computed experimentally. Flow behavior of filament was simulated in 90° bent tube and it was noticed that since plunger action of filament itself is inadequate, therefore aided feeding of filament is needed in order to ensure smooth flow and aesthetic finish of components. The study of effects on mechanical properties, thanks to the introduction of bamboo powder in acrylic polymer (poly-urethane acrylate) revealed that gradual increase in bamboo powder sharply decreases the average stress of composite up to concentration of 2 phr (parts per hundred resin) and slightly increases afterwards. The impact strength however remains approximately the same. The post-processing parameters like surface roughness, dimensional stability and built time can be studied comprehensively for capability to use the composite for tooling applications [82].

9.4   SUMMARY

Intelligent addition of foreign particles in FDM base materials has led to mitigation of inherent downsides of the overall FDM process like surface finish, mechanical property anisotropy [61,79]. Usually, addition of metal or ceramic particles suppresses the tensile properties of the blend mainly because the added particles behave as source of micro voids and stress concentrators. However, polymer blends and typically ternary blends improve the rheology and mechanical property anisotropy of the resulting composites.

9.5   SURFACE ENHANCEMENT THROUGH METAL PLATING

Metal plating could be a promising solution to prevent surface of mold cavity from adhesion with the MIM feedstock, to improve the life of mold and to facilitate release of MIM green-part. The adhesion could result due to direct contact of polymeric mold with the feedstock polymer, at high injection temperature and pressure. In a concept proving trial experiment the authors experienced diffusion of feedstock with the polymer mold cavity. This adhesion prevents the release of the MIM part and degrades the mold life as well. Some studies reported occasional release of a few MIM parts from the polymer molds, yet for reliable and sustainable MIM operation, it is recommended to plate the mold with metallic layer prior to perform MIM [60].

Some studies pertaining to the plating through deposition approach (chemical and vapor) have been deployed to evaluate MIM performance. Jin et al. proposed vapor deposition of di-chloromethane on FDM built PLA molds to improve surface smoothness. This made the substrate elastic and smooth but decreased the mechanical strength due to erosion of the surface [83]. Performance of plated polymer molds for manufacturing of artificial prosthesis, was found comparable to the CNC machined mold or hot-pressed mold. Even at times, polymer coating could be suitable for polymer IM and/or patterns making for casting, yet metal coating is recommended for enhanced and superior performance of the mold [84]. ABS has typical surface angle of 80° so it is classified as hydrophobic, therefore direct action of plating solution is not easy on the ABS. To make it hydrophilic to facilitate its metal coating, the surface needs to be grafted with hydrophilic layer. McCullough et al. [85] treated ABS with acetone solution, followed by grafting with poly ethylene glycol methacrylate, due to which the surface angle was decreased to 40°. Likewise grafting with acrylic acid layer would also render the surface of ABS as hydrophilic [86].

9.5.1 ELECTROLESS METAL PLATING ON POLYMER SUBSTRATES

Plating through direct applications and spraying is usually less adherent and does not accurately conform to the intricate profiles of substrate. Adhesion of metal on the polymer substrate is also a potential challenge. Electro-less plating [87] is potentially a viable solution to plate a polymer while preserving intricate geometric contours and it offers a good adhesion. Contrary to electroplating, where electric current is used to deposit material, electro-less plating deposits the material layers on the polymers by chemical reactions [88]. A process will be defined as electro-less plating if the reduction and deposition of an element is carried out due to chemical reaction of the reducing chemical within solution itself, unaided by external circuit [89]. The first ever coating of metal in the documented history, is believed to be of gold on copper, found in Ur city (presently in Iraq) which ages about 3000–4000 BC [90]. Electro-less plating was first invented by Wurtz in 1844 [91] however, first nickel-tungsten plating was conducted by Brener and Riddkle in 1946 during the Second World War [92,93]. Previously the noble metals were plated using amalgamated solutions however, the associated possible health dangers with the chemical reagents and mercurial heavy metals compelled the researchers to investigate on the new techniques [94]. The electrochemical deposition was then introduced, which latterly transformed into electro-less deposition (ELD). Wurtz in 1844 [95] deployed the reducing characteristics of sodium hypophosphite as a reducing agent for electroless plating and so is said to be the pioneer of the nickel deposition, however, the term of electro-less deposition was introduced by Brenner around 1947 [92].

In applications where surface properties are most important, metal-plated polymer components can be quick and economic solution. Potential applications are in petroleum, automotive, defense and electronic industries [96]. Electro-less deposition (ELD) refers to coating of polymer with metal at low temperature, without application of electric current. Compared with vapor deposition [97], intermediate coating [98] and energy-irradiation [99], ELD is swift, economic and convenient with reduced environmental footprint. Some researchers are focusing to operate the plating bath at lower temperatures and metal plating has successfully been achieved at 40°C [100]. In a study, FDM-made prototype was plated with metal and was successfully used as end-use component [101]. In some of the literature published in past few years [102] the understanding of the electro-less plating on non-metallic substrates has become more illustrated.

9.5.2 ELECTROLESS COPPER PLATING FOR ABS MOLDS

Common FDM polymers for mold making like ABS, polycarbonate, polystyrene, polyethylene and polysulfones can be plated with copper through electroless plating. Typical coating layer thickness is 0.127 μm and the coating is usually uniform. In molds for moderately high-volume demand of MIM parts, electroless plating could be followed by electroplating or patterning. Generally, the process of copper plating is similar to nickel; however, the copper plating solution needs to be alkaline for efficient reducing action of formaldehyde.

The electroless plating process comprises of three main steps; etching, activation and finally metal deposition. The etching is carried out by the action of strong chrome acids [103]. The etchant removes the poly-butadiene and leaves behind the styrene acrylonitrile. The removal of poly-butadiene generates the anchorage points, which dictate the deposition of metals on ABS [104]. Nevertheless, the reduction of chrome ions is challenging and if not reduced, could cause "skip plate" or coating defects. Additionally, the chrome ions (Cr^{6+}) are hazardous due to their carcinogenic nature. The activation after the etching is carried out by the action of palladium or tin salts of halogens [89,105]. Palladium is considered as universal activation catalyst; however, its increasing demand has driven its cost in past a few years. The high cost and toxic nature of the metal could be a challenge for its sustainable use in metal plating industries. Some studies have reported successful use of palladium free metal plating process [106]. Some copper deposition processes have been conducted through coating the substrate with aluminum seeding and using the seed layer for anchoring of copper particles [107]. Typical plating rate is 1–5 µm/h [108]. Typically the peel off strength of 1300 N/cm^2 has been achieved for nickel plating whereas off strength of 2500 N/cm^2 has been achieved for copper plating [109]. However, improved plating processes could further augment the adhesion strength of the coating.

9.6   SUMMARY AND CONCLUSION

If any one or both from the constraints of "low-volume production demands" and "customized demands" exist, FDM-made polymer molds with metal coating could be a promising solution to address such kind of demands. Customized and intricate metallic molds, like those for biomedical prostheses, maintenance parts, and automobiles are usually pricy and require substantial manufacturing lead-time and skills. Typical molds may require a few weeks of extensive machining to qualify necessary required quality control. Moreover, once the part requirement is met or the design is changed, the mold tool is no longer useful. This results in wastage of resources and skills. Nevertheless, the polymeric mold tools are easy and quick to manufacture directly from 3D model, regardless of complexity of geometry and number of features. In case of FDM, manufacturing lead-time and cost of material is seldom dependent on the complexity of mold tool. The potential drawbacks of surface quality and anisotropy in FDM manufacturing could considerably be controlled by optimal selection and enhancements in processing parameters, build material and post-processing. The tailored and enhanced materials have been observed to control anisotropic properties of FDM-made molds. Furthermore, the noticeable advantage of the post-processing through metal plating on the polymer mold is that adhesion of feedstock with mold could be reduced, part release could be facilitated, and eventually, the mold life could be preserved.

9.7   ACKNOWLEDGEMENT

The authors acknowledge the necessary resources provided by Universiti Teknologi PETRONAS (UTP) Malaysia, to conduct this study.

REFERENCES

[1] D. V. Rosato, & M. G. Rosato, *Injection Molding Handbook,* Springer Science & Business Media: NewYork, NY, 2012.

[2] P. Dunne, S. P. Soe, G. Byrne, A. Venus, & A. R. Wheatley, "Some demands on rapid prototypes used as master patterns in rapid tooling for injection moulding," *Journal of Materials Processing Technology,* 150(3): 201–207, 2004.

[3] E. Radstok, "Rapid tooling," *Rapid Prototyping Journal,* 5(4): 164–169, 1999.

[4] F. P. W. Melchels, J. Feijen, & D. W. Grijpma, "A review on stereolithography and its applications in biomedical engineering," *Biomaterials,* 31(24): 6121–6130, 2010.

[5] S. Rahmati, H. Saleem, *Direct Rapid Tooling, in Comprehensive Materials Processing,* Elsevier: Oxford, p. 303–344, 2014.

[6] M. F. Omar, S. Sharif, M. Ibrahim, H. Hehsan, M. N. M. Busari, & M. N. Hafsa, "Evaluation of direct rapid prototyping pattern for investment casting," in *Advanced Materials Research,* (Vol. 463, pp. 226–233). Trans Tech Publications Ltd: Zürich, 2012.

[7] N. J. Karapatis, V. Griethuysen, & R. Glardon, "Direct rapid tooling: a review of current research," *Rapid Prototyping Journal,* 4(2): 77–89, 1998.

[8] A. Bernard, & A. Fischer, "New trends in rapid product development," *CIRP Annals-Manufacturing Technology,* 51(2): 635–652, 2002.

[9] K. S. Boparai, R. Singh, & H. Singh, "Development of rapid tooling using fused deposition modeling: a review," *Rapid Prototyping Journal,* 22(2): 281–299, 2016.

[10] P. L. Kumar, & A. P. Haleem, "Rapid prototyping technology for new product development," *International Journal of Innovative Science, Engineering & Technology,* 3(1): 287–292, 2016.

[11] J. Correa, & P. Ferreira, "Analysis and design for rapid prototyping mechanism using hybrid flexural pivots," *Procedia Manufacturing,* 1: 779–791, 2015.

[12] S. Upcraft, & R. Fletcher, "The rapid prototyping technologies," *Assembly Automation,* 23(4): 318–330, 2003.

[13] Y. Tang, & Y. F. Zhao, "A survey of the design methods for additive manufacturing to improve functional performance," *Rapid Prototyping Journal,* 22(3): 569–590, 2016.

[14] N. A. Waterman, & P. Dickens, "Rapid product development in the USA, Europe and Japan," *World Class Design to Manufacture,* 1(3): 27–36, 1994.

[15] P. He, L. Li, L. Hui, Y. Jianfeng, L. James, & A. Yi, "Compression molding of glass freeform optics using diamond machined silicon mold," *Manufacturing Letters,* 2(2): 17–20, 2014.

[16] P. M. Angelopoulos, S. Michail, & T. Maria, "Functional fillers in composite filaments for fused filament fabrication; a review," *Materials Today: Proceedings,* 37, 4031–4043, 2021.

[17] S. Raut, K. Vijay, S. Jatti, K. K. Nitin, & T. P. Singh, "Investigation of the effect of built orientation on mechanical properties and total cost of FDM parts," *Procedia Materials Science,* 6, 1625–1630, 2014.

[18] C. S. Lee, S. G. Kim, H. J. Kim, & A. Sung-Hoon, "Measurement of anisotropic compressive strength of rapid prototyping parts," *Journal of Materials Processing Technology,* 187, 627–630, 2007.

[19] C. Ziemian, M. Sharma, & S. Ziemian, "Anisotropic mechanical properties of ABS parts fabricated by fused deposition modelling," in *Mechanical Engineering.* InTechOpen: London, UK, 2012.

[20] M. Aslam, F. Ahmad, P. S. M. B. Yusoff, K. Altaf, M. A. Omar, & R. M. German, "Powder injection molding of biocompatible stainless steel biodevices," *Powder Technology,* 295, 84–95, 2016.

[21] J. Abenojar, F. Velasco, J. M. Torralba, J. A. Bas, J. A. Calero, & R. Marce, "Reinforcing 316L stainless steel with intermetallic and carbide particles," *Materials Science and Engineering: A*, 335(1–2): 1–5, 2002.

[22] D. S. Ingole, A. M. Kuthe, TS. B. Hakare, & A. S. Talankar, "Rapid prototyping— a technology transfer approach for development of rapid tooling," *Rapid Prototyping Journal*, 15(4): 280–290, 2009.

[23] S. Olivera, H. B. Muralidhara, K. Venkatesh, K. Gopalakrishna, & C. S. Vivek, "Plating on acrylonitrile—butadiene—styrene (ABS) plastic: a review," *Journal of Materials Science*, 51(8): 3657–3674, 2016.

[24] Z. Zhu, V. G. Dhokia, A. Nassehi, & S. T. Newman, "A review of hybrid manufacturing processes—state of the art and future perspectives," *International Journal of Computer Integrated Manufacturing*, 26(7): 596–615, 2013.

[25] K. Altaf, J. A. Qayyum, A. M. A. Rani, F. Ahmad, P. S. Megat-Yusoff, M. Baharom, & R. M. German, "Performance analysis of enhanced 3D printed polymer molds for metal injection molding process," *Metals*, 8(6): 433, 2018.

[26] S. Masood, & W. Song, "Development of new metal/polymer materials for rapid tooling using fused deposition modelling," *Materials & Design*, 25(7): 587–594, 2004.

[27] ASTM, *Standard Terminology for Additive Manufacturing Technologies F2792 – 12a*. ASTM International: Pennsylvania, USA, pp. 1–3, 2012.

[28] I. D. Gibson, & B. Stucker, *Additive Manufacturing Technologies: 3D Printing, Rapid Prototyping, and Direct Digital Manufacturing*, Springer; New York, NY, 2014.

[29] C. K. Chua, & K. F. Leong, "Rapid prototyping: principles and applications," *World Scientific*, 1, 2003.

[30] P. Jain, & A. Kuthe, "Feasibility study of manufacturing using rapid prototyping: FDM approach," *Procedia Engineering*, 63: 4–11, 2013.

[31] M. Chhabra, & R. Singh, "Rapid casting solutions: a review," *Rapid Prototyping Journal*, 17(5): 328–350, 2011.

[32] L. C. Hieu, N. Zlatov, J. Vander Sloten, E. Bohez, L. Khanh, P. H. Binh, & Y. Toshev, "Medical rapid prototyping applications and methods," *Assembly Automation*, 25(4): 284–292, 2005.

[33] R. S. Singh, & P. Kapoor, "Investigating the surface roughness of implant prepared by combining fused deposition modeling and investment casting," *Proceedings of the Institution of Mechanical Engineers, Part E: Journal of Process Mechanical Engineering*, 230(5): 403–410, 2016.

[34] K. Subburaj, & B. Ravi, "Computer aided rapid tooling process selection and manufacturability evaluation for injection mold development," *Computers in Industry*, 59(2): 262–276, 2008.

[35] K. Altaf, J. A. Qayyum, A. M. M. Rani, F. Ahmad, P. S. M. Megat-Yusoff, M. Baharom, A. R. A. Aziz, M. Jahanzaib, & R. M. German, "Performance analysis of enhanced 3d printed polymer molds for metal injection molding process," *Metals*, 8(6): 433, 2018.

[36] C. L. Perez, "Analysis of the surface roughness and dimensional accuracy capability of fused deposition modelling processes," *International Journal of Production Research*, 40(12): 2865–2881, 2002.

[37] A. K. Sood, R. K. Ohdar, & S. S. Mahapatra, "Experimental investigation and empirical modelling of FDM process for compressive strength improvement," *Journal of Advanced Research*, 3(1): 81–90, 2012.

[38] A. Boschetto, & L. Bottini, "Surface Characterization in Fused Deposition Modeling," in *3D Printing: Breakthroughs in Research and Practice*, IGI Global: Hershey, PA, pp. 22–47, 2017.

[39] M. Taufik, & P.K. Jain, "Role of build orientation in layered manufacturing: a review," *International Journal of Manufacturing Technology and Management*, 27(1–3): 47–73, 2013.

[40] K. Thrimurthulu, P.M. Pandey, & N.V. Reddy, "Optimum part deposition orientation in fused deposition modeling," *International Journal of Machine Tools and Manufacture*, 44(6): 585–594, 2004.

[41] E. Sabourin, S. A. Houser, & J. Helge Bøhn, "Accurate exterior, fast interior layered manufacturing," *Rapid Prototyping Journal*, 3(2): 44–52, 1997.

[42] E. L. Melgoza, G. Vallicrosa, L. Serenó, J. Ciurana, & C. A. Rodríguez, "Rapid tooling using 3D printing system for manufacturing of customized tracheal stent," *Rapid Prototyping Journal*, 20(1): 2–12, 2014.

[43] R. S. Anitha, Arunachalam, & P. Radhakrishnan, "Critical parameters influencing the quality of prototypes in fused deposition modelling," *Journal of Materials Processing Technology*, 118(1–3): 385–388, 2001.

[44] P. K. Gurrala, & S. P. Regalla, "DOE Based Parametric Study of Volumetric Change of FDM Parts," *Procedia Materials Science*, 6: 354–360, 2014.

[45] A. K. Sood, R. Ohdar, & S. Mahapatra, "Improving dimensional accuracy of fused deposition modelling processed part using grey Taguchi method," *Materials & Design*, 30(10): 4243–4252, 2009.

[46] R. Singh, "Some investigations for small-sized product fabrication with FDM for plastic components," *Rapid Prototyping Journal*, 19(1): 58–63, 2013.

[47] R. J. Singh, J. Singh, & S. Singh, "Investigation for dimensional accuracy of AMC prepared by FDM assisted investment casting using nylon-6 waste based reinforced filament," *Measurement*, 78: 253–259, 2016.

[48] J. S. Chohan, R. Singh, & K. S. Boparai, "Mathematical modelling of surface roughness for vapour processing of ABS parts fabricated with fused deposition modelling," *Journal of Manufacturing Processes*, 24: 161–169, 2016.

[49] J. W. Carr, & C. Feger, "Ultraprecision machining of polymers," *Precision Engineering*, 15(4): 221–237, 1993.

[50] P. M. Pandey, N. V. Reddy, & S. G. Dhande, "Improvement of surface finish by staircase machining in fused deposition modeling," *Journal of materials processing technology*, 132(1): 323–331, 2003.

[51] D. Espalin, J. A. Ramirez, F. Medina, & R. Wicker, "Multi-material, multi-technology FDM: exploring build process variations," *Rapid Prototyping Journal*, 20(3): 236–244, 2014.

[52] M. Lay, N. L. N. Thajudin, Z. A. A. Hamid, A. Rusli, M. K. Abdullah, & R. K. Shuib, "Comparison of physical and mechanical properties of PLA, ABS and nylon 6 fabricated using fused deposition modeling and injection molding," *Composites Part B: Engineering*, 176, 107341, 2019.

[53] A. K. Sood, R. K. Ohdar, & S. S. Mahapatra, "Parametric appraisal of mechanical property of fused deposition modelling processed parts," *Materials & Design*, 31(1): 287–295, 2010.

[54] S. Kumar, & J. P. Kruth, "Composites by rapid prototyping technology," *Materials & Design*, 31(2): 850–856, 2010.

[55] O. C. Ivanova, Williams, & T. Campbell, "Additive manufacturing (AM) and nanotechnology: promises and challenges," *Rapid Prototyping Journal*, 19(5): 353–364, 2013.

[56] D. R. Askeland, & W. J. Wright, *Science and Engineering of Materials,* Cengage Learning: Stanford, CA, 2015.

[57] W. Brostow, R. Chiu, I. M. Kalogeras, & A. Vassilikou-Dova, "Prediction of glass transition temperatures: Binary blends and copolymers," *Materials Letters*, 62(17–18): 3152–3155, 2008.

[58] J. Duesing, "Law of averages," *Leonardo*, 30(5): 397–397, 1997.
[59] D. Roberson, C. M. Shemelya, E. MacDonald, & R. Wicker, "Expanding the applicability of FDM-type technologies through materials development," *Rapid Prototyping Journal*, 21(2): 137–143, 2015.
[60] J. Rajaguru, M. Duke, & C. Au, "Development of rapid tooling by rapid prototyping technology and electroless nickel plating for low-volume production of plastic parts," *The International Journal of Advanced Manufacturing Technology*, 78(1–4): 31–40, 2015.
[61] N. Sa'ude, S. H. Masood, M. Nikzad, M. Ibrahim, & M. H. I. Ibrahim, "Dynamic mechanical properties of copper-ABS composites for FDM feedstock," *International Journal of Engineering Research and Applications*, 3(3): 1257–1263, 2013.
[62] N. Mostafa, H. M. Syed, S. Igor, & G. Andrew, "A study of melt flow analysis of an ABS-Iron composite in fused deposition modelling process," *Tsinghua Science & Technology*, 14(Supplement 1): 29–37, 2009.
[63] M. Nikzad, S. H. Masood, & I. Sbarski, "Thermo-mechanical properties of a highly filled polymeric composites for Fused Deposition Modeling," *Materials & Design*, 32(6): 3448–3456, 2011.
[64] D. A. Roberson, R. R. Carmen, & M. Piñon, "Evaluation of 3D Printable Sustainable Composites," *26th Annual Solid Freeform Fabrication Symposium*, Austin, Texas, USA: University of Texas, 2015.
[65] A. R. Torrado, C. M. Shemelya, J. D. English, Y. Lin, R. B. Wicker, & D. A. Roberson, "Characterizing the effect of additives to ABS on the mechanical property anisotropy of specimens fabricated by material extrusion 3D printing," *Additive Manufacturing*, 6: 16–29, 2015.
[66] A. R. T. Perez, D. A. Roberson, & R. B. Wicker, "Fracture surface analysis of 3D-printed tensile specimens of novel ABS-based materials," *Journal of Failure Analysis and Prevention*, 14(3): 343–353, 2014.
[67] C. R. Rocha, A. R. T. Perez, D. A. Roberson, C. M. Shemelya, E. MacDonald, & R. B. Wicker, "Novel ABS-based binary and ternary polymer blends for material extrusion 3D printing," *Journal of Materials Research*, 29(17): 1859–1866, 2014.
[68] D. K. Dimitrov, Schreve, & N. De Beer, "Advances in three dimensional printing-state of the art and future perspectives," *Rapid Prototyping Journal*, 12(3): 136–147, 2006.
[69] L. Kuentz, A. Salem, M. Singh, M. C. Hailbig, & J. A. Salem, "Additive manufacturing and characterization of Polylactic Acid (PLA) composites containing metal reinforcements," *International Conference and Expo on Advanced Ceramics and Composites, NASA*, The American Ceramic Society OH, USA, 2016.
[70] S. E. Moving, Montgomery, & D. Somos, "Rapid Tooling via Stereolithography," *RTE Journal*, 2006. www.rtejournal.de
[71] Gy. Falk, J.G. Kovacs, & K. Toth, "Development results in rapid tooling," *International Polymer Science and Technology*, 32(2): 365–368, 2005.
[72] A. Bagsik, V. Schöppner, & E. Klemp, "FDM part quality manufactured with Ultem* 9085," *14th International Scientific Conference on Polymeric Materials*, 2010. www.stratasys.com
[73] L. Novakova-Marcincinova, & I. Kuric, "Basic and advanced materials for fused deposition modeling rapid prototyping technology," *Industrial & Manufacturing Engineering*, 11(1): 24–27, 2012.
[74] M. L. Shofner, K. Lozano, F. J. Rodríguez-Macías, & E. V. Barrera, "Nanofiber-reinforced polymers prepared by fused deposition modeling," *Journal of applied polymer science*, 89(11): 3081–3090, 2003.
[75] R. W. Gray IV, D. G. Baird, & J. Helge Bøhn, "Effects of processing conditions on short TLCP fiber reinforced FDM parts," *Rapid Prototyping Journal*, 4(1): 14–25, 1998.

[76] S. Singh, & R. Singh, "Development of functionally graded material by fused deposition modelling assisted investment casting," *Journal of Manufacturing Processes*, 24: 38–45, 2016.

[77] S. Onagoruwa, S. Bose, & A. Bandyopadhyay, *Fused Deposition of Ceramics (FDC) and Composites*, Pro SFF, TX, p. 224–31, 2001.

[78] G. Wu, N. A. Langrana, R. Sadanji, & S. Danforth, "Solid freeform fabrication of metal components using fused deposition of metals," *Materials & Design*, 23(1): 97–105, 2002.

[79] A. Bellini, S. Guceri, & M. Bertoldi, "Liquefier dynamics in fused deposition," *Journal of Manufacturing Science and Engineering*, 126(2): 237–246, 2004.

[80] S. R. Stewart, J. E. Wentz, & J. T. Allison, "Experimental and computational fluid dynamic analysis of melt flow behavior in fused deposition modelling of Poly (lactic) Acid," *ASME 2015 International Mechanical Engineering Congress and Exposition*, American Society of Mechanical Engineers, 2015.

[81] N. M. I. A. Isa, N. Sa'ude, M. Ibrahim, S. M. Hamid, & K. Kamarudin, "A study on melt flow index on copper-abs for fused deposition modeling (FDM) feedstock," in *Applied Mechanics and Materials* (Vol. 773, pp. 8–12). Trans Tech Publications Ltd: Zürich, 2015.

[82] S. H. Chiu, I. Ivan, C. L. Wu, K. T. Chen, S. T. Wicaksono, & H. Takagi, "Mechanical properties of urethane diacrylate/bamboo powder composite fabricated by rapid prototyping system," *Rapid Prototyping Journal*, 22(4): 676–683, 2016.

[83] Y. Jin, Y. Wan, B. Zhang, & Z. Liu, "Modeling of the chemical finishing process for polylactic acid parts in fused deposition modeling and investigation of its tensile properties," *Journal of Materials Processing Technology*, 240: 233–239, 2017.

[84] J. Noble, K. Walczak, & D. Dornfeld, "Rapid tooling injection molded prototypes: a case study in artificial photosynthesis technology," *Procedia CIRP*, 14: 251–256, 2014.

[85] E. J. McCullough, & V. K. Yadavalli, "Surface modification of fused deposition modeling ABS to enable rapid prototyping of biomedical microdevices," *Journal of Materials Processing Technology*, 213(6): 947–954, 2013.

[86] A. Garcia, T. Berthelot, P. Viel, A. Mesnage, P. Jégou, F. Nekelson, & S. Palacin, "ABS polymer electroless plating through a one-step poly (acrylic acid) covalent grafting," *ACS Applied Materials & Interfaces*, 2(4): 1177–1183, 2010.

[87] Y. Shacham-Diamand, Y. Sverdlov, S. Friedberg, & A. Yaverboim, "Electroless plating and printing technologies," in *Nanomaterials for 2D and 3D Printing*, Wiley: New York, 2017.

[88] Y. W. Laima Luo, J. Li, Y. Zheng, "Preparation of nickel-coated tungsten carbide powders by room temperature ultrasonic-assisted electroless plating," *Surface and Coatings Technology*, 206(6): 1091–1095, 2011.

[89] J. Sudagar, J. Lian, & W. Sha, "Electroless nickel, alloy, composite and nano coatings—A critical review," *Journal of Alloys and Compounds*, 571: 183–204, 2013.

[90] L. B. Hunt, "The oldest metallurgical handbook," *Gold Bulletin*, 9(1): 24–31, 1976.

[91] G. O. Mallory, & J. B. Hajdu, *Electroless plating: fundamentals and applications*. William Andrew, AESF: Orlando, FL, 1990.

[92] A. Brenner, & G. E. Riddell, "Deposition of nickel and cobalt by chemical reduction," *Journal of Research of the National Bureau of Standards*, 39: 385–395, 1947.

[93] G. Gutzeit, & P. Talmey, "Chemical nickel plating processes and baths therefor," *Google Patents*, 1958. www.patents.google.com.

[94] E. Darque-Ceretti, & M. Aucouturier, "Gilding for matter decoration and sublimation. A brief history of the artisanal technical know-how," in *1st International Conference on Innovation in Art Research and Technology*, HAL Open Science: Evora, 2013.

[95] A. Wurtz, "On copper hydride (Translated from French)," *Comptes rendus de l'Académie des Sciences*, 18: 702–704, 1844.

[96] X. Tang, M. Cao, C. Bi, L. Yan, & B. Zhang, "Research on a new surface activation process for electroless plating on ABS plastic," *Materials Letters*, 62(6–7): 1089–1091, 2008.

[97] W. Su, L. Yao, F. Yang, P. Li, J. Chen, & L. Liang, "Electroless plating of copper on surface-modified glass substrate," *Applied Surface Science*, 257(18): 8067–8071, 2011.

[98] Y. Liao, B. Cao, W. Wang, C. Zhang, L. Wu, D. & R. Jin, "A facile method for preparing highly conductive and reflective surface-silvered polyimide films," *Applied Surface Science*, 255(19): 8207–8212, 2009.

[99] Y. Wang, C. Bian, & X. Jing, "Adhesion improvement of electroless copper plating on phenolic resin matrix composite through a tin-free sensitization process," *Applied Surface Science*, 271: 303–310, 2013.

[100] Z. Shu, & X. Wang, "Environment-friendly Pd free surface activation technics for ABS surface," *Applied Surface Science*, 258(14): 5328–5331, 2012.

[101] Y. Ding, H. Lan, J. Hong, & D. Wu, "An integrated manufacturing system for rapid tooling based on rapid prototyping," *Robotics and computer-integrated manufacturing*, 20(4): 281–288, 2004.

[102] B. J. Sherwood, "Serendipity produces a radically new plating process: Electroless nickel," *Metal Finishing*, 106(6): 92–94, 2008.

[103] N. Inagaki, & H. Kimura, "Electroless copper plating on acrylonitrile butadiene styrene material surfaces without chromic acid etching and a palladium catalyst," *Journal of Applied Polymer Science*, 111(2): 1034–1044, 2009.

[104] L. Di, B. Liu, J. Song, D. Shan, & D. A. Yang, "Effect of chemical etching on the Cu/Ni metallization of poly (ether ether ketone)/carbon fiber composites," *Applied Surface Science*, 257(9): 4272–4277, 2011.

[105] W. Zhao, Q. Ma, L. Li, X. Li, & Z. Wang, "Surface modification of ABS by photocatalytic treatment for electroless copper plating," *Journal of Adhesion Science and Technology*, 28(5): 499–511, 2014.

[106] X. Tang, C. Bi, C. Han, & B. Zhang, "A new palladium-free surface activation process for Ni electroless plating on ABS plastic," *Materials Letters*, 63(11): 840–842, 2009.

[107] B. Karagoz, O. Sirkecioglu, & N. Bicak, "Surface rejuvenation for multilayer metal deposition on polymer microspheres via self-seeded electroless plating," *Applied Surface Science*, 285: 395–402, 2013.

[108] J. R. Henry, "Electroless (autocatalytic) plating," *Metal Finishing*, 100: 409–420, 2002.

[109] S. Arai, & T. Kanazawa, "Electroless deposition and evaluation of Cu/multiwalled carbon nanotube composite films on acrylonitrile butadiene styrene resin," *Surface and Coatings Technology*, 254: 224–229, 2014.

10 Process Design for Hot Forging via Finite Element Analysis Considering Reverse Engineering

Omar Youssef, Cuneyt Boz and Volkan Esat

CONTENTS

This book chapter aims to address a fundamental metal forming process design keeping the rationale of employing reverse engineering in mind. The manuscript presents the steps of a methodology leading to modelling the somewhat complicated process of hot forging. The computational methodology may not only be used to

DOI: 10.1201/9781003220985-10

successfully construct hot forging dies that can produce near-net-shape or even net-shape parts using reverse engineering, but also utilized to figure out various process parameters or material properties which help understand and/or analyze any burdensome hot forging operation. I sincerely believe that the content can significantly contribute to your event as well as to the book to be published.

10.1 INTRODUCTION

Forging is one of the oldest manufacturing techniques in the world. Long before the Industrial Revolution, it was known that many metals are easier to work when they are hot. Hot forging is an excellent method of inducing both strength and toughness into solid metal and shape it using a large force, control with great precision developing the maximum impact strength, fatigue resistance, and superior internal integrity. As time passed, the use of metals became more extensive and more advanced techniques for metalworking have emerged [1]. Metal forming processes constitute one of the largest pillars of manufacturing for shaping metallic materials into useful products of engineering with the inclusion of dimensional accuracy and desired mechanical characteristic properties [2]. There are many metalworking branches and techniques that are widely used today. Some of these processes include bulk metal forming, such as rolling, extrusion, and forging, which are usually carried out around certain temperatures to obtain specific material properties and easy workability.

Since the beginning of earliest manufacturing, forging has inferred great skill and labour as an experience-oriented technology, largely controlled by the know-how passed on by generations of trial and error. Today, forging is applied on a vast scale. The operation can be fast and for mass production, or a slow and delicate one. Forgings can be utilized in almost all industries; however, are prevalent in the automotive, aerospace, marine, and agricultural sectors, and for producing many general industrial equipment. Despite being cost effective, expensive dies are justified due to typical high production rates and less scrap yield. On the other hand, various early failures may occur.

In the process of hot forging, die life and conditions are limited by wear, mechanical fatigue, and plastic deformation. To observe these failure modes and their adverse effects in the context of this research study, analyses were conducted through process modelling under different conditions. The analyses were carried out by using the computer aided three-dimensional finite analysis software, MSC. Marc. The properties of the material in the simulations were taken from the database of another manufacturing FEA software, Simufact. While analyzing the hot forging process, results were obtained by simulating a series of successive stages of the operation in order to achieve the most accurate and realistic output. As an initial stage, rigid dies and cold work experiments were modelled to build upon the fundamentals of forging. After that, by defining properties as functions of temperature, hot forging experiments were numerically conducted. In this way, the differences between hot forging and cold forging processes were highlighted. Then, hot forging analyses were set initially on rigid dies. For this purpose, deformable-deformable contact models were constructed with different parameters to observe the processes more realistically. Friction and temperature were the main process parameters that were gradually applied and varied in the simulations. The effects of these parameters on the material properties as well as on the internal stresses were observed. Changing the friction and temperature affected the flow stress in the workpiece, and subsequently, the strains and stresses on the punch and die. These simulations were performed not only for the open die, but also for the closed die operations. Sensitivity analyses were also performed within the scope of this project in order to determine the minimum number of finite elements that produced the converged estimations.

Some researchers applied reverse engineering concepts in metal forming. Chenot et al. [3] reported on the rheological parameters starting with a known constitutive equation coupled with finite element method (FEM) and an optimizing algorithm that uses the least-square method to minimize an objective function which contains the material parameters. Pietrzyk et al. [4] evaluated rheological parameters and friction coefficients in metal forming processes by the means of an inverse method where they identified variables in the flow equation, friction model, and internal variable model's dislocation density in one set of ring compression test. They found that it is more ideal to determine rheological parameters and friction coefficients from one combined test as the analysis of results depends on an initial friction factor. Zhiliang et al. [5] invented a new technique combining compression tests and FEM (C-FEM) to improve the flow stress by taking interface friction into account and minimizing it as a target function defined in the load-stroke curves. A flow chart describing reverse engineering in metal forming is shown in Figure 10.1.

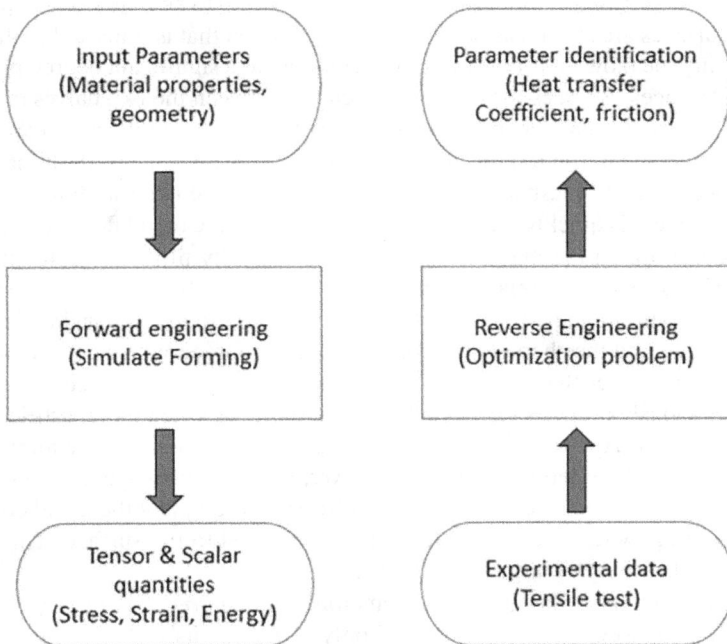

FIGURE 10.1   Forward and Reverse Engineering Concepts in Metal Forming.

10.2 LITERATURE SURVEY

In this section, types of hot forging processes, die failure modes, and process parameters are covered within the context of reverse engineering in order to highlight some of the most crucial concepts in hot forging.

10.2.1 HOT FORGING PROCESSES

Forging begins with an ingot, blooms, or billets of metal, which can weigh anything up to 70 tons [6]. The process will not only change the shape of the ingot but also its internal formation. This process refines its grain structure removing any voids, forging any undesirable elements along the whole axis of the ingot, imparting a fibrous structure, which in turn strengthens the material. Hot forging can be classified into three main types according to Groover [7].

Open die forging in which a forging hammer applies an impact load imparting its gravitational/pressurized energy to the workpiece compressing it between two flat surfaces with the work piece free to expand in the lateral directions. This generally requires multiple hits, and has less accuracy, but gives the freedom to move in multiple directions. Forging press steadily shapes the metal and typically has a higher forging force. Another part of this family is ring rolling which increases the diameter of the billet after a hole has been punched into it. This process can be used in upsetting, cogging, and edging. Skill of the human operator is a factor in the success of these operations.

Impression die forging is considered by many sources to be a closed die operation. The die surfaces are shaped as or contain an impression that is imparted to the workpiece during the blow, thus constraining metal flow to a significant degree providing tighter tolerances. It forces the metal into a cavity between the two halves of the die. Drop forging is an ideal way of manufacturing all sorts of complicated shapes while allowing excess material to flow outside the die impression to form flash that should be trimmed off later. Flash serves an important role as the material flows out of the impression. It cools quickly increasing the resistance of the metal flowing outside the die and constraining the material to remain in the cavity filling the intricate parts. Flash thickness is closely related to wear and fatigue life. Impression dies can forge everything from small levers and gears to intricate crankshafts by varying the weight of the hammer and the shape of the die. The metal is forged with a continuous force applied slowly, rather than repeated hammer blows. The force is much more powerful. It is obviously a faster process. The high cost of the machine is justified by the increased output. Another benefit is that the dies can be more easily aligned with each other providing better accuracy. However, the big advantage of pressing over drop forging is that the pieces are worked uniformly throughout the metal compared to drop forging, where the impact tends to be confined to the surface. This can be significant when forging large components.

Flashless forging is the third type, sometimes called *precision forging*. It is considered to be a closed die operation, and only differs by its lack of flash. At a first glance, this operation seems favorable; however, it imposes some requirements. Most importantly, the initial work volume must be within a close tolerance of the space in the die cavity. If the starting volume is too big, the press will try to squeeze it using high pressures that may cause damage to the die. If the blank is too small, the cavity will not be filled, and the product will be defective.

Isothermal forging is a special type of hot forging in which the dies are heated to the same temperature of the work piece virtually eliminating temperature-difference-induced stresses or thermal fatigue. This process is expensive and reserved for alloys difficult to forge such as titanium [6]. These alloys are usually developed to have high strength at elevated temperatures; hence, the difficulty forging them, generally requiring low strain rates as this will have a strong bearing on the flow stress [7]. Die wear is naturally more accelerated, and highly evident in such operations.

10.2.2 DIE FAILURE MODES

Failure of dies is a complex and time-dependent process. Several factors come into play like die material and hardness, work-metal composition, forging temperature, condition of the work metal at forging surfaces, and workpiece design. Failure of dies can be classified into three main modes; namely, wear, fatigue fracture, and plastic deformation. According to Schey [8], die wear is responsible for nearly 70% of die failures.

Wear is a slow removal of material off a component caused by erosion, corrosion, abrasion, and adhesion. Abrasion resulting from friction is deemed the highest in die wear. It has been proven that greater resistance to abrasion and adhesion is closely related to higher strength and hardness of the die. Therefore, in hot forging, the die steel should have a high hot hardness and retain this hardness over extended periods of time [6].

Abrasion by hard oxide such as scale is a major concern, and it happens when a generally soft material is removed by a harder one from another surface, this is called two-point abrasion. Three-point abrasion is when loose particles trapped between the surfaces cause the "ploughing" [6]. Adhesion is when there is a metal-to-metal contact between the surface asperities, and one metal shears generally the workpiece, and sticks to the die [8].

Contact time, temperature, sliding velocity, friction factor, surface finish and lubricant all have a bearing on hot forging, which makes it a complicated process to model and handle.

Higher contact times of the die with the elevated workpiece temperature will result in wear for two reasons, one being higher temperature decreases the hardness of the die, and second is that as the temperature of the die increases and the billet cools down, the flow stress increases demanding higher pressure from the die [6]. In addition, if the press has low stiffness, then it has more contact time due to elastic deflection during loading, thus, more wear according to Kesavapandian et al. [9]. Moreover, Dahl et al. state that if the temperature keeps fluctuating due to higher contact times and cooling, thermal fatigue might be a problem [10]. This is especially pertinent to hydraulic presses as the contact time is higher. However, forging hammers are more susceptible to fatigue as they impart their energy in a faster manner. Toughness is also a function of strength and ductility; it usually increases with increasing temperature according as reported by Altan et al. [6].

In order to decrease the contact time, one might be tempted to increase the press speed. However, sliding velocity between the die and the billet is an important factor for wear, and the press speed influences the sliding velocity. High sliding velocities increase the strain rate which in turn increases the flow stress resulting in wear, and decreased fatigue life [6]. In addition, higher sliding velocities generate heat in the interface that lowers the hardness of the die; thus, more wear according to Tulsyan et al. [11].

Die design parameters such as flash geometry, fillet radii, draft angles, and die face contact area all influence die wear and fatigue life. According to Aston et al. [12], it has been found that higher flash thickness result in low contact stresses; therefore, decreasing wear, and increasing fatigue life. However, the opposite happens if flash-metal escape rate is high as the die exerts higher pressures to compensate for the high flash-metal escape rates.

Rougher surfaces result in more wear according to Tulsyan et al. [11]. This is expected as the rougher a surface is, the more the contact occurs for a lesser number of asperities, causing higher loads per asperity. If a surface is too "clean"; then, adhesive wear is a concern unless coatings are used [6].

The lubricant type influences the contact pressure, temperature, and heat transfer. A good lubricant can minimise heat transfer between the billet and the die, increasing the thermal fatigue life, and decrease wear by lowering the interface pressure. Too much lubricant can cause a build-up of particles in the die cavities causing premature failure according to Schey and Dahl et al. [8,13]. Moreover, the higher the billet temperature, the more the scale on the workpiece surface [11].

In a research study, a FEM model corroborated the empirical data that die wear increases with decreasing hardness. However, wear decreases with increasing corner radius and friction factor. At first glance, it does not seem to be intuitive that increasing friction factor would decrease wear, as mentioned before, increasing the friction factor would increase contact pressures, hence wear. However, in some cases, the friction factor decreased the sliding length which was more detrimental according to Dahl et al. [13]. Many researchers have tried to estimate die wear in hot forging, and the results have been similar to experiments however, the exact magnitudes were difficult to estimate according to Dahl et al., and Lee and Jou [13,14].

Archard's model is one of the oldest analytical wear models. It aims to predict abrasive and adhesive wear. It is expressed in Equation (1) below [6]:

$$V = \frac{K.P.S}{H} \tag{10.1}$$

Where V is the wear volume, K is the experimental wear coefficient, P is the normal pressure, S is the sliding length, and H is the hardness of the softer material. Thus, as can be seen, that the higher the load, the greater the wear; the higher the hardness, the lower the wear. Also, the longer it slides, the more wear. It appears reasonable; however, the problem rises with the K constant. K have been determined for many die/workpiece material combinations. Bay [15] claims that adhesive wear, K, varies based on the material pair, and the lubricant used.

Other developments were carried out on Archard's model by Felder and Montagut [16] in order to reach a simpler form for the abrasive wear model which proved useful in plane stress problems. An alternative energy approach by Stahlberg and Hallstrom [17] accounts for the energy supplied by the machine.

Fatigue is one of the important failure factors that, according to a common material science definition, is explained as crack initiation and crack propagation in material is caused by repetitive loading [18]. Fatigue failure in hot forging process can occur due to high thermal and mechanical strain. Damages caused by fatigue can occur on the die surface as well as in the subsurface [19]. This condition causes considerable variations on die life.

Mechanical fatigue cracking occurs due to applying cyclic forces below material's static yield strength. The characteristic die life range expectation in hot forging is 10,000 to 50,000 cycles. However, due to mechanical fatigue, die failures can be seen earlier than expected [20]. Like mechanical fatigue, thermal fatigue is also extremely important in hot working. It is defined as crack formation on the die due to repetitive temperature changes [19]. Altering temperature causes thermal expansion on dies in hot working process, and these temperature differences cause stress and strain values to change. The continuous repetition of this process results in the formation of cracks on or in the die, followed by growth and propagation of these cracks.

Failures of fatigue usually occur in three main phases that follow each other. These are; formation of crack, growth of crack, and final fracture, respectively. In the first stage, cracks initiate in microscopic scale within die material, which can merge in high stress fields in part or areas with high void density. Then, these micro-scale cracks are enlarged in stage two, and finally, when cracks reach a critical point, final fracture occurs. Stage one takes longer compared to other stages. As cracks progress in stages, shear stress in the material also increases [21].

During forging process, dies and punches are exposed to extensive forces. Due to this situation, internal pressure and stress intensity on die material become very high. Higher die stresses due to applied considerable forces cause early fatigue failure. Optimal forging forces lead to obtain optimum die life, and in addition, decrease cost of production. To be able to adjust the optimal force, geometry and dimensions of preform have a great influence on die life in terms of fatigue [22].

Another parameter to consider in hot forging is thermal fatigue which is related to the temperature of the die. Due to heat transfer between dies and workpiece during the repetitive action, die temperature rises owing to the high temperature difference between the two, therefore a hot-cold cycle occurs above the die temperature. As a result of this, softening of the material of die, tempering on material, and decreasing strength can be observed [20]. In various research work on forging temperature, stabilizing the die temperature is not an easy task unless a lubricant or cooling system is provided. Additionally, elevated temperatures affect die life adversely. Also, die temperature can be affected by several other parameters such as billet temperature, lubricant type, and environmental factors [20].

Fatigue failure in hot forging process depends on parameters such as temperature, material mechanical properties, and forging force. Moreover, there are different other variables that affect these parameters. Controlling the entire set of factors affected by fatigue is a challenge that can be achieved through FEA simulation.

10.2.3 Process Parameters

In hot forging process design, many parameters have an impact on failure process. The relations between these parameters, which would have positive or negative effects on die failure, are highly complicated. To obtain maximum die or punch life, it is necessary to understand the trends and tendencies [6]. The prominent parameters in hot forging process design are discussed in this section.

Temperature and thermal events in hot working process are critical factors as they are highly complicated. In hot working, temperature distribution primary depends on the workpiece temperature, die temperature, heat generation, friction, and heat transfer rate [6]. In hot forging process, temperature affects material's flow stress. Accordingly, it dictates the power to be applied, and the amount of energy expended. High temperatures in the process cause decrease in strain-hardening; therefore, for the billet, high temperatures may be regarded as an advantage in decreasing the power requirement, whereas, it has an adverse effect on the die [23].

Strain rate is another important factor for workability, which is defined as materials capability of deform without fracture. Strain rate and workability are some of the key features when determining the material properties [6]. Flow stress of material is stated as a function of strain, strain rate, and temperature in various models developed for cold and hot work. A most basic and highly common form to describe flow stress in hot work is given in Equation (10.2) below.

$$\sigma_f = C\mu^m \tag{10.2}$$

where, $C = f(\varepsilon, T)$. C and m variables are studied and reported by many researchers. According to these, increasing the temperature decreases the flow stress, while increasing the strain rate increases the flow stress [6].

Friction plays a crucial role in forging as expected. Metal flow takes place as a consequence of applying pressure from the die onto the billet. The friction at the interface influences how this process progresses [6]. In the absence of friction, pressure is expected to be uniformly distributed along contact area; however, in the case of friction, shear stresses associated with the friction force form, causing barreling due to the non-uniformly distributed pressure. The stresses are higher than those in the frictionless case that build up in the center and decrease with increasing diameter.

There are four main lubrication methods according to Altan et al. [6] that influence metal forming depicted up in the Stribeck curve in terms of lubricant viscosity, sliding velocity, and normal pressure [6].

Friction models quantify friction, which is important in predicting forging pressures. There are two famous models: namely, Coulomb's law, and shear friction law. Coulomb's law is not valid at all pressures for the fact that shear stress cannot exceed the shear strength; therefore, shear friction law has been developed [6]. Shear friction law can be expressed as follows:

$$\tau = f\,\bar{\sigma} = 0.577m\,\bar{\sigma} = mk \tag{10.3}$$

Where friction factor m ranges from 0 to 1, and k is the shear strength of the material. In hot forming of steels, m roughly ranges from around 0.2 to 0.4 as opposed to cold forming, in which it ranges from 0.05 to 0.15 [6].

10.2.4 REVERSE ENGINEERING

Reverse engineering usually denotes a process in which the original component's data is being captured for a multitude of reasons; redesigning and optimization purposes instead of creating the solid model using conventional CAD or analyzing a failed component by reconstructing its physical damaged parts. Furthermore, it allows for the possibility of recreating parts which are no longer in the market. Somrack [24] utilized reverse engineering on a worn component to retrace the geometry of its die cavity to perform finite element analysis for redesigning purposes. The technique employed a microscribe robotic arm which is a CMM method that digitized the physical features of the component. This technique can be used with several CAD systems; however, in this case, SolidWorks 3-D CAD system was used in tandem with RevWorks reverse engineering add-on in SolidWorks. A finite element analysis was performed, where it was found that the component was overstressed, yielding two different solutions. In a study by Ramnath et al. [25], reverse engineering was used to optimize a crankshaft's fatigue performance of three different materials commonly used in automotive crankshafts. Laser scanner device was used to generate the cloud points in this non-contact 3-D CMM. CAD was performed using CATIA, and dynamic analysis was conducted and verified on ANSYS. In a study by Esat et al. [26], both geometric reverse engineering and functional reverse engineering were explored in two different machine tools. Bench-type drilling machine and metal/wood cutting band saw were disassembled and measured then modelled on a CAD system (Autodesk Inventor). Finite element, sustainability, and cost analyses were performed on the most critical part and then an evaluation matrix based on different criteria was used to find the optimum solution to improve that critical component. Figure 10.2 illustrates the process flow diagram of the reverse engineering concept in obtaining either process parameters or identifying geometry.

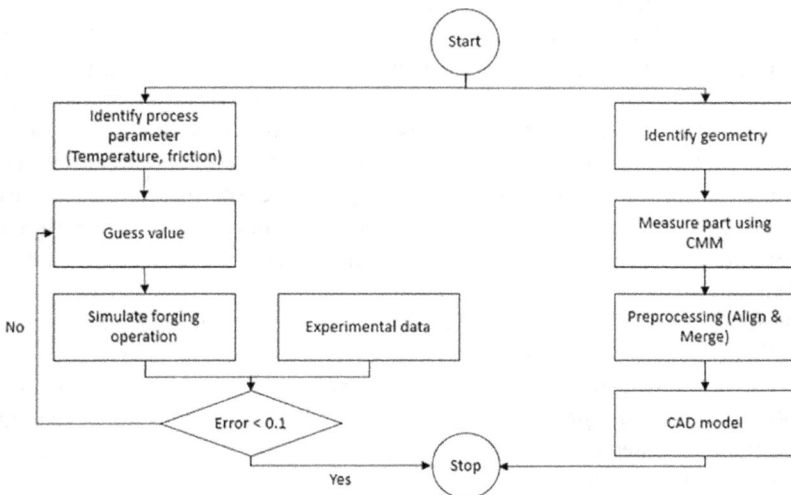

FIGURE 10.2 Flow Process of Reverse Engineering Concept in Metal Forming.

10.3   THEORY AND METHODOLOGY

10.3.1   MODELLING: FINITE ELEMENT ANALYSIS

Hot forging is inherently a non-linear process with large strains incurred. Development of finite element method in the late 1970s was inadequate to help simulate hot forging mainly because automatic meshing was not available according to Ngaile et al. [27]. However, several advances have been achieved since then in the computational technology, and today, finite element analysis is an asset, and an indispensable tool for researchers and engineers. The main objective of FEA in forging can be summarized in three points. First is developing a satisfactory die design by preventing flow-induced defects such as laps and predicting temperature so that other detrimental factors such as die wear can be minimized. Second is improving part quality by improving grain flow and material yield. Third is predicting forging loads and stresses induced in the die so that premature failure can be avoided, and appropriate machines can be selected for a given application; hence being significantly cost effective.

According to Howson et al. and Oh [28, 29] modelling of closed die forging has been used in aerospace industry for decades. In the early stages, FEM helped die design engineers simulate metal flow, and possible defect formation. Then, results in the form of contour state variables such as effective stress and strain, and temperature could be visualized. Further advancements are looking into microstructure modelling with metallurgical aspects available like grain size, and precipitation predicted with increasing accuracy [30,31]. The main goal is enhancing the product and improving the performance of the components by better understanding the parameters controlling it.

Depending on its geometric complexity, a forging process can be modelled as a two-dimensional, axisymmetric or plane-strain, or a three-dimensional problem. A good starting position is modelling in an axisymmetric fashion using quadrilateral elements and applying the necessary boundary/symmetry conditions. If done properly, the simulation is almost exact to its three-dimensional counterpart.

Hot forging is characterized by large plastic deformation. To simulate accurate metal flow for large strain, and strain rates, the flow curve is generally obtained from a carefully extrapolated compression test. Furthermore, the flow curve is defined as a function of strain, strain rate and temperature [6]. Moreover, material properties that relate heat transfer to deformation such as thermal conductivity, and heat capacity of the workpiece are defined. Young's modulus and Poisson's ratio are important elastic material parameters to analyse die stress [6].

The starting mesh is defined, and can be refined in critical areas; however, as the simulation proceeds, mesh distortion becomes a concern. This phenomenon can be handled by the auto-generation of a new mesh taking the updated geometry into consideration, called as *adaptive remeshing*. A common example that can be anticipated in time and eliminated by FE is lap formation through changing workpiece geometry or die geometry, or both.

In a study conducted by Brucelle et al. [32], FEA is used to increase tool life, which is further validated by experimental results. They investigated the thermomechanical stresses on the dies in the hot extrusion of an automotive component. The dies were subjected to mechanical loadings as it punches the workpiece and thermal cyclic ones due to contact with the workpiece. The thermal stresses were found to comprise 75% of the die stress. As a result, the dies could only withstand 500 cycles due to thermal cracks. It is well known that changing the geometry of the dies has no bearing on thermal stresses. A better design could be achieved by modifying other parameters like temperature of the billet and punching speed of the die. It was decided to decrease the temperature of the billet, although the flow stress would increase, the punch could handle the loads due to increased flow stress. This process was modelled using rigid dies to save computing time which in the end was changed to elastic, and the stresses were then interpolated into the die to show the equivalent stress, principal stresses, and temperature distribution. An optimal die design was achieved that resulted in 30% decrease in die stresses and increased tool life.

In this chapter, the first model presented is based on open die forging. There are three stages to be followed in adequately modelling hot forging with FEM. The first stage is modelling cold work with rigid dies. The second stage is adding the thermal analysis. The third stage is changing the rigid geometric dies with deformable bodies to accurately simulate hot forging. After a successful FEM model is constructed, a final step should be validation. Sensitivity analysis is a crucial part of modelling, which means validating the mesh either by doing a convergence plot which keeps refining the mesh till it converges to a selected result, or by manually calculating the error of two or three consecutive mesh sizes till it becomes insignificant, effectively eliminating the mesh size, i.e., the number of elements, as a limiting factor on the results predicted.

10.3.2 ASSUMPTIONS AND LIMITATIONS OF FEA IN METAL FORMING

Finite element analysis is favored in metal forming as it provides a systematic way for a better insight into critical design parameters, and reducing time and money spent to produce a new product. Commercial software such as MSC Marc offer a non-linear FEA solution to accurately simulate material behavior in many complex problems especially in contact analysis. However, the user is required to have a basic understanding of modelling and FEM. Furthermore, the user should know about the process parameters used in the analysis, and appropriate meshing. The main assumption which is a valid simplification for axisymmetric problems is the initial/boundary conditions which are idealized at the beginning of each simulation. Nevertheless, it should be noted that axisymmetric analysis does not determine buckling since, every radial deformation is the same (due to the boundary condition) which is generally not the case in buckling. Therefore, a separate 3D model should be investigated for buckling. Another assumption is that the plasticity model used is inherent to the material chosen which can be found in Marc's database. However, Marc offers several constitutive analytical models or even user specified models for more freedom. As a primer, the first model uses a rigid plastic material model which is popular for it is numerically robust and ignores strain hardening which means less computing time. However, all further models use elastic-plastic formulations as specified in the relevant sections. It should also be noted that coefficients such as friction or heat transfer are extremely critical as discussed in detail in the forthcoming sections.

10.3.3 FUNDAMENTAL PROBLEMS AND DEVELOPED MODELS

Axisymmetric modelling is preferred due to the nature of the selected forging problem. Modelling elastic-plastic problems is the first step. Modelling efforts start with building an axisymmetric punch problem. In this basic model, a geometric punch is moving downward in the vertical direction deforming and indenting the workpiece which is held against a die. This elementary model is used to apply all the necessary steps for conducting an elastic-plastic contact problem which are explained in detail in the oncoming sections when introducing the first model developed.

The second fundamental problem is thermal contact. To simulate hot forging, another sample axisymmetric problem is constructed as a key to figure out how thermal/structural analysis is carried out, especially when both entities in contact are deformable. Basically, there is a pipe enclosed in a housing which is heated from the inside. The pipe and housing initially do not touch; when the pipe heats up, it expands, and at a certain moment, it comes in contact with the housing. This model helps understand the basic steps of thermal/structural modelling.

10.3.3.1 First Model: Cold Work with Rigid Punch and Die

The first model developed simulates cold work of a typical carbon steel for elastic-plastic problems. As an open die forging problem, it is chosen as an axisymmetric cold work upsetting which is governed by Hollomon's law defined as $\sigma_f = K\varepsilon^n$. Depicted in Figure 10.3(a), the geometry and mesh are generated. Symmetry boundary condition was applied on the left vertical edge of the workpiece. Young's modulus and Poisson's ratio are entered as $E = 210$ GPa, and $v = 0.3$, respectively. Figure 10.3(a) and (b) shows the axisymmetric and 360 revolved views.

Furthermore, plastic true stress-strain curve is entered which governs the plastic behavior of the steel as tabulated in Table 10.1.

Punch motion is set so that the full stroke happens in a few seconds, soon after which sudden retraction of the punch is facilitated. Moreover, contact interactions

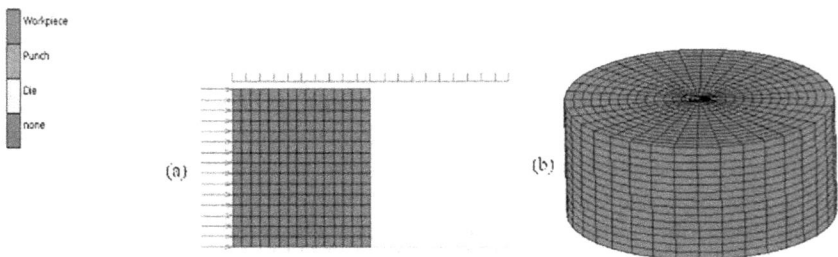

FIGURE 10.3   (a) Axisymmetric and (b) 360 Revolved Views of First Model.

TABLE 10.1
True Plastic Stress vs Strain Data.

μ_{pl}	σ [MPa]
0	300
0.1	370
0.2	400
0.3	420
0.4	430
0.6	440

are defined. The contact is defined for the workpiece to be between both the punch and the die. However, for the release stage, it is only defined between the workpiece and the die. The findings for all of the developed models are provided and discussed in the next main section, *results and discussion*.

10.3.3.2   Second Model: Hot Work with Rigid Punch and Die

The second model is constructed to incorporate hot forging, introducing thermal analysis to make the model one step closer to reality. The punch and die are still geometric rigid entities. The key point in the thermal analysis of bulk metal forming, which is governed by elastic-plastic isotropic constitutive equations, is determining how a material deforms with increasing temperature. Therefore, in addition to the elastic modulus which varies with temperature; thermal conductivity, specific heat capacity, and thermal expansion coefficients are also functions of temperature. The values are taken from Simufact materials database [33] and adapted into the model so that they match the unit system chosen for the plasticity data in Table 10.1. The plasticity properties are taken from Marc's database [34] and are in the same form as Table 10.1 used in the first model, where stresses and strains are all true quantities, ignoring the elastic part. Therefore, if the user would input their own values, the plastic strain is:

$$\varepsilon_{plastic} = \varepsilon_{total} - \frac{\sigma_{true}}{E} \qquad (10.4)$$

Furthermore, a major change in the dimensions of the billet is made to be eligible to use the heat transfer coefficient reported by Chang and Bramley [35]. This parameter is arguably one of the most influential parameters and highly elusive to investigate. The billet is 25 mm in diameter and 37.5 mm in height. The heat transfer coefficient was measured by thermocouples. The reported value is 0.365 [kW/m².C] when sitting on the die pre-forging. This value accurately corresponds to the value predicted in their FE model. The problem arises when forging takes place. The value rises significantly; therefore, an average value taken over 10 increments is proposed to be 7.79 [kW/m².C]. Emissivity is also taken to be 0.7. The convection coefficient which is defined for the outer edge (right side) is reported to be 0.043 [kW/m².C]. An inverse algorithm with an iterative approach is used to predict the interface heat transfer coefficient using both the calculated and experimental temperatures by applying the least squares method combined with an inverse algorithm to obtain an error 0.2%. Table 10.2 summarizes the heat transfer coefficients used. Figure 10.4 shows the new billet in the second model with contact definitions.

10.3.3.3 Third Model: Hot Work with Deformable Punch and Die

The third, and arguably the most realistic, model comprises of a deformable punch and die. This allows for a structural and thermal analysis of the dies, as well. A hard material is chosen as a die and punch material to withstand the high stresses without much strain. An enhanced H13 tool steel [36] is used with its mechanical and thermal properties plugged in as functions of temperature.

In order to move the punch, a straight line defined as a rigid geometric entity and specified as a contact body glued to the punch. It is then used to push the punch downwards to upset the workpiece using a defined velocity.

The contact is carefully defined since node to segment control is the method of choice. The workpiece, which is more finely meshed, is assigned as the slave entity

---

TABLE 10.2

Summary of Heat Transfer Rate Coefficients.

Heat transfer rate coefficient [kW/m².C]

Rest-on die	0.365
During forging	7.79
Convection	0.043

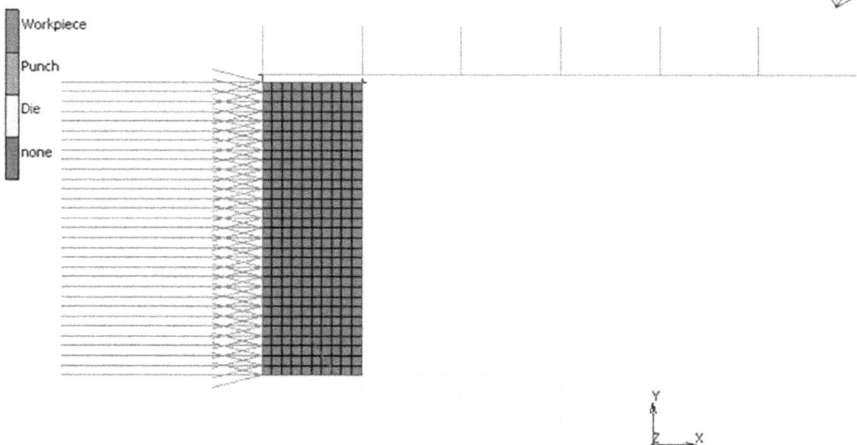

FIGURE 10.4 Second Model Contact Entities.

while the punch and die which are stiffer are assigned as the master entities. This is because Marc allows master to penetrate into the slave. If the slave is densely meshed, the element size of slave is small enough to allow the nodes to effectively prevent penetration into their sphere of influence. Segment to segment methods is a relatively new method introduced to Marc which eliminates the master/slave concept. It appears to be promising; however, further investigation is needed to distinguish any practical differences.

The workpiece is initially meshed 10 x 30 using four-node quadrilateral elements (Element no 10) and the dies are meshed using 3-node triangular elements (Element no 2). The die and the punch are more densely meshed near the surface of contact shown in Figure 10.5. Global adaptive meshing is utilized on the workpiece as excessive deformation is present throughout the forging process. The criteria for remeshing is fairly simple, although Marc provides lots of tools to control the remeshing. The main remeshing criteria are *frequency* and *immediate* options. Frequency is controlled by either increments or time steps and is assigned as five increments in the model as a typical reasonable value. Immediate option is not used readily; however, is only utilized when element distortion, penetration, or angle deviation from undeformed is 30° (90° being the reference). Minimum edge length is specified to be 0.5 mm for all results; however, sensitivity analysis is performed first to determine the best mesh criterion with associated dimensions. These criteria have served well with reasonable computing time.

The chosen problem requires that the billet is reduced to 65% of its height. It is also customary and advised in FEA of metal forming to use *updated Lagrangian*

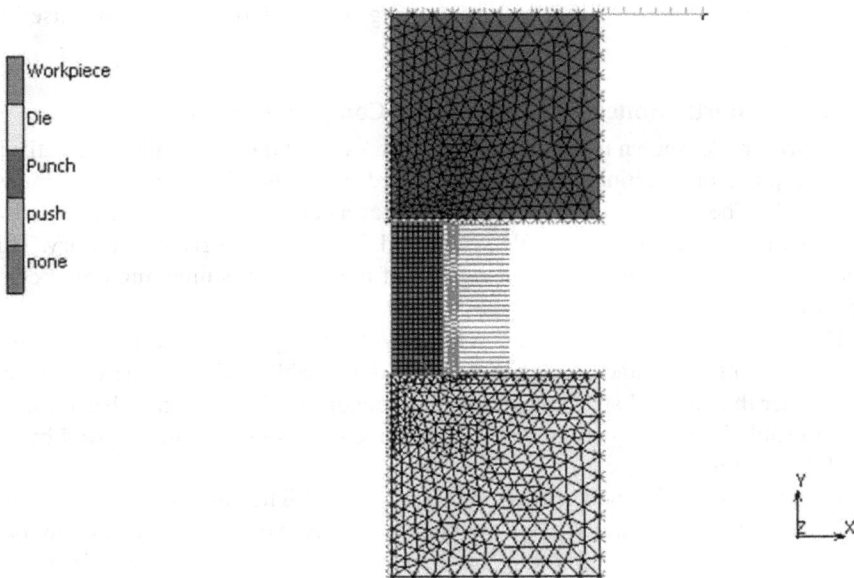

FIGURE 10.5   Third Model Contact Entities.

procedure instead of *total Lagrangian* for stress and strain measures, although both are expected to lead to theoretically and numerically similar results. The choice of one over another is dictated by such things as convenience of modelling the physics of the problem, rezoning requirements, and integration of constitutive equations. Updated lagrange is naturally formulated in terms of Cauchy (true) stress and logarithmic strain since the current configuration is the reference configuration. It is used in conjunction with *multiplicative decomposition* which allows larger increments of strain to be used, with greater accuracy and better convergence. The main difference between them is that total Lagrangian evaluates all integrals with respect to the initial undeformed configuration while the latter evaluates with respect to the last completed iteration of the current increment.

MSC.Marc offers several friction models; however, the most widely used by far are Coulomb's friction model and Tresca's shear model. Coulomb's friction model operates on the principle that the normal force is the major factor affecting the friction process, so the friction coefficient, μ is defined as the ratio of tangential stress over normal stress; thus, increasing the normal stress should increase shear stress, which is very intuitive and as expected. On the other hand, Tresca's shear model states that the friction factor, m, is a fraction of the yield stress in shear, σ_{ys}, and is defined for a range between $0 < m < 1$. Both are reported to have similar results. However, it should be noted that when using Coulomb's friction model in metal forming, it doesn't make sense for the normal stresses to be so high because the shear stress cannot exceed the shear strength of the material. This is where Tresca's model shine allowing proper representation of the frictional shear stress. Coulomb's friction model is more prevalent in cold working, while the latter in hot working. Both friction models are used in the analyses.

10.3.3.4   Fourth Model: Hot Forging of a Complex Workpiece

The fourth model shown in Figure 10.6 simulates closed die hot forging. A relatively intricate piece of machine element is produced with flash that should be trimmed afterwards. The machine element can be used as a cam, for example. The die is left straight whereas the punch's profile is changed. This choice is rather arbitrary. This model is created to demonstrate the rigidity of the model in simulating rather complex shapes and operations.

The main difference is that the symmetry boundary condition is now created using a symmetry contact body touching all deformable bodies at the center (left side) rather than fixed displacement boundary condition. The reason is that since the punch profile is not a square any more, the punch's motion is being skewed by the fixed displacement.

The final shape obtained is shown in Figure 10.7(a). The circular outer ring is the flash and should be trimmed after the operation. Reverse engineering can be used to trace back the profile of the dies in order to replicate the exact shape by taking measurements of the final product especially if the shapes are quite intricate with the caveat that certain modifications for thermal expansion should be considered.

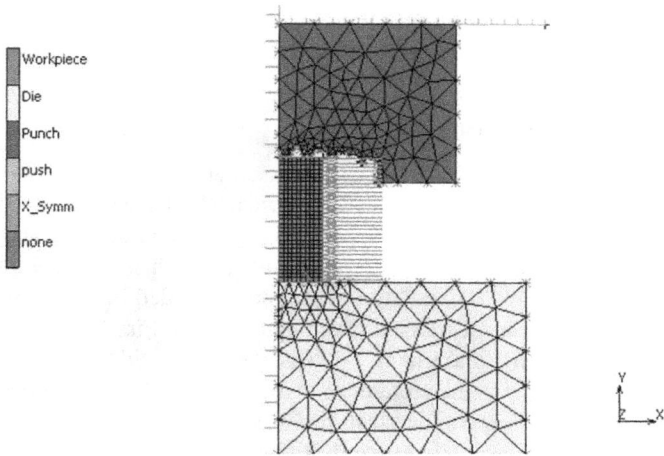

FIGURE 10.6 Fourth Model Contact Entities.

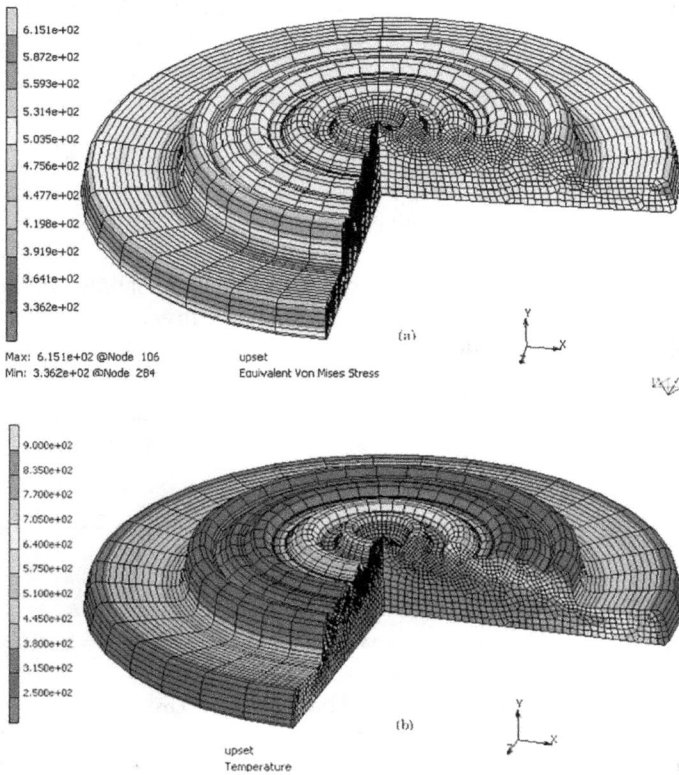

FIGURE 10.7 (a) 270° Revolved View of von Mises Stress Distribution on/in the Closed Die Hot Forged Workpiece, and (b) Temperature Distribution within the Closed Die Hot Forged Workpiece.

Figure 10.7(b) depicts a representative temperature distribution within the workpiece soon after the operation.

10.3.4 SENSITIVITY ANALYSIS

Mesh size is one of the oldest problems since the emergence of FEM. How to ensure adequate mesh size to yield near-exact numerical results without sacrificing too much computing time has always been a challenging undertaking. The consensus is that finer mesh produces more accurate results, which is expected since more parts of the body is discretized, and therefore, more detailed information is carried. Sensitivity analysis is a must to ensure accurate results. First, a metric should be chosen to be analyzed; the maximum von Mises stresses within the workpiece is a good selection, which is also the preference in this work. Maximum von Mises stress at the fully loaded state are compared in this study. There are several ways to conduct it depending on the problem at hand. Reducing the element size (i.e., increasing the number of elements) is the easiest mesh refinement strategy. This method has its drawbacks as certain areas might need more reduction (stress concentrators). Increasing the element order adds inherent accuracy as each element contains more nodes; however, this requires high computing power. Global adaptive meshing is the method of choice for the cases in this study. The mesher generates new mesh at the critical regions using the criteria put forth by the user (minimum element length, penetration, strain %, etc.). This works well for the problems in this study since there are large strains all over the body. An initial 1.25 mm edge length is chosen, and then using smaller edge lengths, maximum von Mises stress is calculated at the fully loaded state. Table 10.3 shows the values with the errors in each trial for the open die hot forging model.

It is observed that the von Mises stress increases incrementally with decreasing element edge length. However, the change is minimal with each increment. The error between the initially chosen edge length and 0.5 mm which is the value that the mesh size appears to be converging to, is less than 1%, which is quite reasonable. This means that the initial guess is a reasonable one with much less computing time than the one corresponding to 0.5 mm.

TABLE 10.3
Open Die Sensitivity Analysis for Workpiece.

Number of elements	Element edge length [mm]	von Mises fully loaded [MPa]	Error	Error final-initial
424	1.25	519.0		0.954%
	1	520.8	0.346%	
	0.8	521.9	0.211%	
2463	0.5	524.0	0.401%	
	0.4	523.2	0.153%	

Sensitivity analysis for no friction case @ $T = 910°C$

10.4 RESULTS AND DISCUSSION

10.4.1 PREDICTIONS: FIRST MODEL

This section presents the results obtained from the first model's cold work simulation and will not go into details of interpreting the results as cold work is not the main aim of this research; however, some remarks are made when necessary. No friction, $\mu = 0.05$, and $\mu = 0.1$ cases are compared at the end. Figure 10.8 shows the

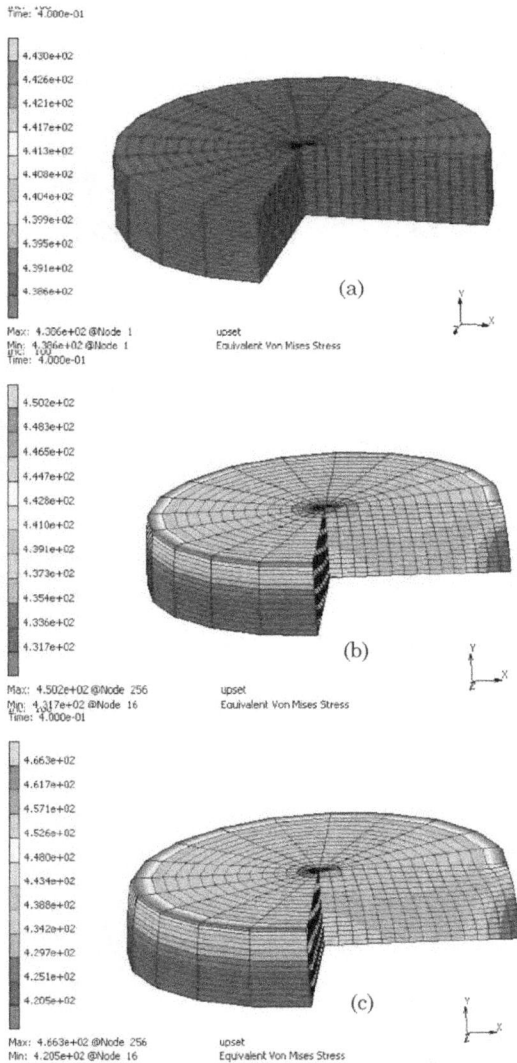

FIGURE 10.8 First Model: von Mises Stresses at the Fully Loaded State for (a) No Friction, (b) $\mu = 0.05$, and (c) $\mu = 0.1$.

fully loaded state of the cut-out workpiece for the *no friction*, $\mu = 0.05$, and $\mu = 0.1$ cases.

Furthermore, a plot of μ vs maximum von Mises stresses reveals that increasing friction increases the stress. As the punch keeps moving downward, the friction at the interface resists this outward motion inducing the non-uniform stress distribution. Figure 10.9 shows the μ vs σ_{vM}. It is apparent in this case that the friction effect increases von Mises stress in an almost linear fashion.

10.4.2 Predictions: Second Model

The effect of friction can be seen on the hot forged workpiece by comparing the maximum von Mises stresses at the fully loaded state in Figure 10.10. An increasing

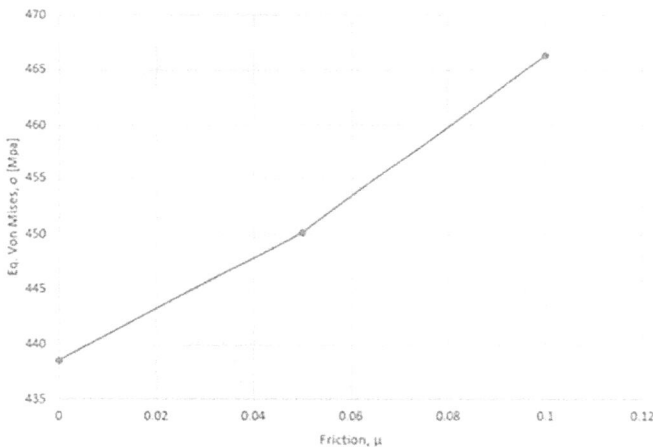

FIGURE 10.9   First Model: Friction Coefficient μ vs Maximum von Mises Stress, σ $_{vM}$.

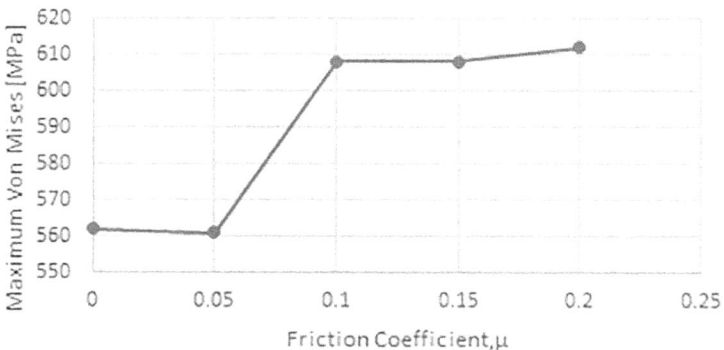

FIGURE 10.10   Second Model: Friction Coefficient μ versus Maximum von Mises Stress, σ $_{vM}$.

trend is observed with increasing friction. Increasing friction coefficient dramatically increases maximum von Mises stress with a sharp increase between 0.05 and 0.1 values of μ.

10.4.3 PREDICTIONS: THIRD MODEL

One of the most important results that is gathered from the all-deformable hot forging model for the purpose of increasing the reliability of the model is the change of flow stress with increasing friction coefficient. Figure 10.11 demonstrates the effect of friction for both fully loaded, and unloaded states in terms of maximum von Mises stress. An increasing linear trend is found with increasing friction. This is in good agreement with the literature, and is reasonable to expect the maximum von Mises stress to increase for both fully loaded and unloaded state.

Another important result is the effect of initial workpiece temperature on flow stress. Figure 10.12 shows the effect of temperature for both fully loaded state and unloaded one on the maximum von Mises stress. It can be observed that for the fully loaded state the trend is almost horizontal which can be explained by the balancing out of strain hardening effect of the upsetting action and the thermal softening with increasing temperature. However, the unloaded state shows more fluctuation which can be attributed to mechanical stress relaxation after elastic recovery, and thermal stresses continuing to shape the stress profile after the punch is withdrawn.

A comparison between the maximum von Mises stress using rigid and deformable punch at the fully loaded state is provided in Figure 10.13. It is observed that the deformable punch is more sensitive to the increase in friction coefficient than the rigid punch, overtaking it and setting higher von Mises stress values for the workpiece. This result shows that for the presence of ideally low interface friction,

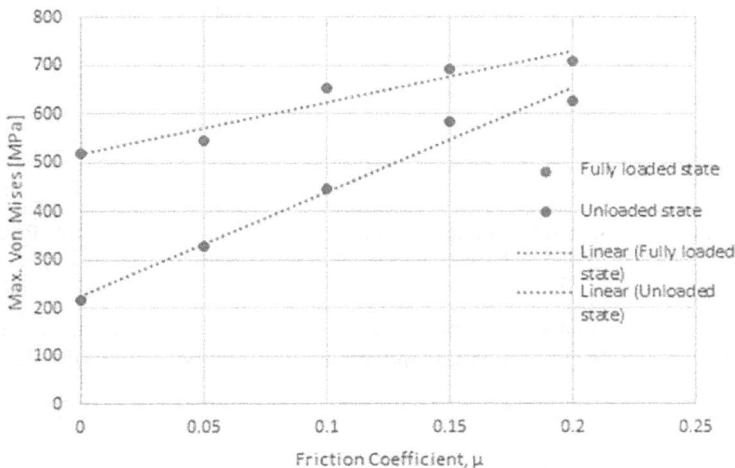

FIGURE 10.11 Third Model: Friction Coefficient μ versus Maximum von Mises Stress, σ $_{vM}$ for Fully Loaded and Unloaded States.

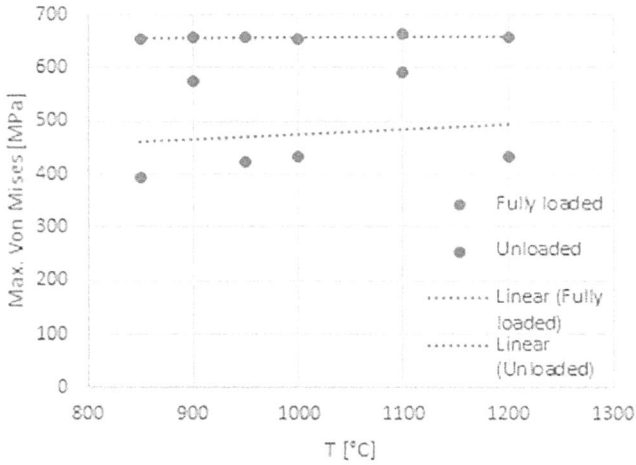

FIGURE 10.12 Third Model: Temperature versus Maximum von Mises Stress, σ vM for Fully Loaded and Unloaded States.

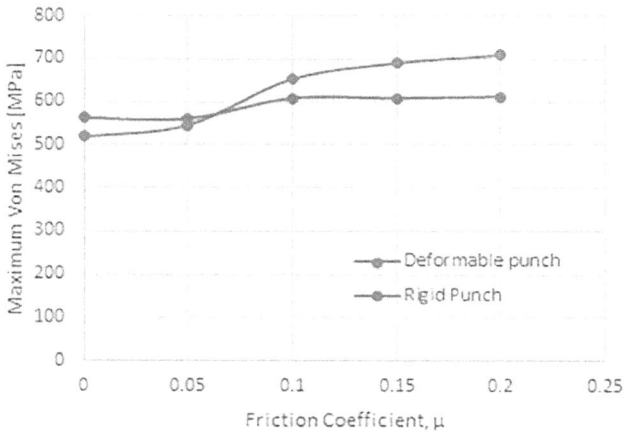

FIGURE 10.13 Comparison of Friction Effect between Rigid and Deformable Punch.

the difference is not that significant; however, for high friction conditions, the difference starts to become consequential, and attention should be given to the deformable punch.

10.4.4 EXPERIMENTAL VALIDATION

This work can be validated by referring to a close study conducted by Chang and Bramley [35] to determine the interface heat transfer coefficient (IHTC) using an inverse algorithm. Initial guess of the IHTC is made, then temperature is extracted and compared to the experimental temperature. Iterations continued until values were found to be within a reasonable margin of error, ranging between around 700 to 900 °C for the initial forging stages and around 300 to 500 °C for the rest-on-die stage at various selected workpiece locations. Final temperature predictions between our work and theirs are found to be in good agreement.

This work aims to make an incremental contribution to the computational research on hot forging. Building upon the available literature and methodologies, it presents a thorough guide on how to model and simulate hot forging through a commercial FEA software such as MSC.Marc, addressing its assumptions, limitations, and recommended settings. The scientific knowledge on how the process parameters such as forging temperatures, heat transfer and friction coefficients, and their variation with respect to each other as well as time, are highlighted. The findings and predictions can be utilized ensure accurate results that resembles the physical phenomenon.

10.5 CONCLUSION

In conclusion, a comprehensive literature survey is conducted on hot forging, examining its principles, various forms, and critical process variables. Furthermore, wear and failure mechanisms are discussed. Reverse engineering concepts are introduced in the scope of metal forming processes, and previous studies on the inverse analysis is reported. Essential process parameters are introduced, and their effects are explained in detail. Finite element analyses' role in investigating bulk metal forming is highlighted. Advancements made by researchers are reported including difficulties and shortcomings. This work briefly reviews previous literature while presenting the required data and process parameters for a complete guide on how to model and simulate open/closed die hot forging in MSC. Marc using reverse engineering techniques.

Some fundamental problems on how to model axisymmetric elastic-plastic materials and conduct thermal/structural analyses are addressed using the FEA medium MSC. Marc. First model is created by modelling the dies as rigid and based on cold forging. The second model including thermal analysis along with structural is the first step towards hot forging. Third model marks deformable—deformable contact within the context of hot work. Fourth model simulates closed die hot forging with a more complex shape. Equivalent von Mises stress and temperature distributions are collected and interpreted. The importance of determining the interface heat transfer coefficient, and how it drastically changes during forging is addressed in theory and methodology. It has been found that increasing friction coefficient will increase the maximum von Mises stresses both in fully loaded or unloaded states. Furthermore, increasing temperature is found to fluctuate the maximum von Mises stress distributions but in an increasing linear trend as the thermal stress is coupled with the structural stress. This phenomenon is especially obvious for the unloaded state in Figure 10.12 where the thermal stresses remain in addition to structural residual stresses after the punch is retracted.

This study aimed to put forward the importance of determining the process parameters for hot forging in forward or reverse engineering, and the role and strength of FEA in fulfilling these objectives. Material libraries offered by the software helped achieve accurate simulations of hot forging. A brief summary of different methods and criteria to choose from for conducting sensitivity analysis to ensure accurate mesh size is provided. Finally, it can be concluded that longevity of hot forging dies could be extended, and near-net-shape or even net-shape products can be successfully manufactured by following the reverse engineering techniques mentioned. It should be noted that part of this work is compared against an experimental and computational study done by C. Chang and A. Bramley [35], warranting further investigations to find an appropriate IHTC for the forging period via modelling the IHTC in terms of other process parameters. Further testing of the models for various hot forging operations may be carried out for a broader validation effort.

REFERENCES

[1] H. Osborne, *XXth Century Sheet Metal Worker—A Modern Treatise on Modern Sheet Metal Work*, 3rd ed. Chicago, IL: American Artisan, 1910.

[2] Y.V.R.K. Prasad, K.P. Rao and S. Sasidhar, eds., *Hot Working Guide: A Compendium of Processing Maps*. Novelty, OH: ASM International Materials Park, 2015.

[3] J.L. Chenot, E. Massoni and J.L. Fourment, "Inverse problems in finite element simulation of metal forming processes," *Engineering Computations*, vol. 13, no. 2/3/4, pp. 190–225, 1996.

[4] M. Pietrzyk and J. Jedrzejewski, "Identification of parameters in the history dependent constitutive model for steels," *CIRP Annals*, vol. 50, no. 1, pp. 161–164, 2001.

[5] L. Xinbo, Z. Fubao and Z. Zhiliang, "Determination of metal material flow stress by the method of C-FEM," *Journal of Materials Processing Technology*, vol. 120, no. 1–3, pp. 144–150, 2002.

[6] T. Altan, S. Oh and H. Gegel, *Cold and Hot Forging: Fundamentals and Applications*, 1st ed. Metals Park, OH: American Society for Metals, 1983.

[7] M. Groover, *Fundamentals of Modern Manufacturing: Materials, Processes, and Systems*, 7th ed. Hoboken, NJ: John Wiley & Sons, 2020.

[8] J. Schey and M. Shaw, "Tribology in metalworking: Friction, lubrication and wear," *Journal of Tribology*, vol. 106, no. 1, pp. 174–174, 1984.

[9] G. Kesavapandian, G. Ngaile and T. Altan, "Die wear in precision hot forging-effect of process parameters and predictive models," *Engineering Research Center for Net Shape Manufacturing*, 2001 [Online]. Available: https://ercnsm.osu.edu/

[10] C. Dahl, V. Vazquez and T. Altan, "Effect of process parameters on die life and die failure in precision forging," Report No. PF/ERC/NSM-98-R-15, Engineering Research Center for Net Shape Manufacturing, April 1998.

[11] R. Tulsyan, R. Shivpuri and T. Altan, "Investigation of die wear in extrusion and forging of exhaust valves," Report No. ERC/NSM-B-93–28, Engineering Research Center for Net Shape Manufacturing, August 1993.

[12] J. Aston and E. Barry, "A further consideration of factors affecting the life of drop forging dies," *The Journal of the Iron and Steel Institute*, vol. 210, no. 7, pp. 520–526, July 1969.

[13] C. Dahl, V. Vazquez and T. Altan, "Analysis and prediction of die wear in precision forging operations," Report No. PF/ERC/NSM-99-R-21, Engineering Research Center for Net Shape Manufacturing, May 1999.

[14] R.S. Lee and J. Jou, "Application of numerical simulation for wear analysis of warm forging die," *Journal of Materials Processing Technology*, vol. 140, no. 1–3, pp. 43–48, 2003. doi:10.1016/s0924-0136(03)00723-4.

[15] N. Bay, *Class Notes*. Lyngby: Technical University of Denmark, 2002.

[16] S. Babu, D. Ribeiro and R. Shivpuri, *Materials and Surface Engineering for Precision Forging Dies*. Columbus, OH: The Ohio State University, 1999.

[17] F. Gorczyca, *Application of Metal Cutting Theory*. New York: Industrial Press, 1987.

[18] J. Schijve, "Fatigue of structures and materials in the 20th century and the state of the art," *International Journal of Fatigue*, vol. 25, no. 8, pp. 679–700, 2003. Available: www.sciencedirect.com/science/article/abs/pii/S0142112303000513?via%3Dihub [Accessed 14 April 2021].

[19] S. Chander and V. Chawla, "Failure of hot forging dies—An updated perspective," *Materials Today: Proceedings*, vol. 4, no. 2, pp. 1147–1157, 2017. Available: www.science direct.com/science/article/pii/S2214785317301311 [Accessed 14 April 2021].

[20] D. D'Addona and D. Antonelli, "Application of numerical simulation for the estimation of die life after repeated hot forging work cycles," *Procedia CIRP*, vol. 79, pp. 632–637, 2019. Available: www.sciencedirect.com/science/article/pii/S2212827119301799 [Accessed 14 April 2021].

[21] J. Feritz and H. Dolares, "Preliminary research for the development of a hot forging die life prediction model," *Etd.ohiolink.edu*, 2021 [Online]. Available: https://etd.ohiolink. edu/apexprod/rws_etd/send_file/send?accession=ohiou1102695461&disposition=inl ine [Accessed 14 April 2021].

[22] V. Pandya and P. George, "Analysis of die stress and forging force for DIN 1.2714 die material during closed die forging of anchor shackle," *Materials Today: Proceedings*, p. 7, 2021. Available: https://doi.org/10.1016/j.matpr.2021.01.121 [Accessed 15 April 2021].

[23] M. Milutinović, D. Vilotić and D. Movrin, "Precision forging—tool concepts and process design," *Journal for Technology of Plasticity*, vol. 33, no. 1–2, pp. 86–87, 2008. Available: www.semanticscholar.org/paper/Number-1-2-PRECISION-FORGING-%E2 %80%93-TOOL-CONCEPTS-AND-Milutinovi-Viloti/980df0ef73d23824d3aed9d413 81d794c1de830d [Accessed 16 April 2021].

[24] A. Somrack, "Finite element analysis of a forging die whose cavity was obtained by reverse engineering techniques," *Citeseerx.ist.psu.edu*, 2021 [Online]. Available: http:// citeseerx.ist.psu.edu/viewdoc/summary?doi=10.1.1.624.7926 [Accessed 3 September 2021].

[25] B. Vijaya Ramnath et al., "Implementation of reverse engineering for crankshaft manufacturing industry," *Materials Today: Proceedings*, vol. 5, no. 1, pp. 994–999, 2018.

[26] K. Buyuktas, M. Suhaib, W. Joyia, K. Karimov, I. Khan and V. Esat, "Machine tool improvement through reverse engineering and computational analysis with an emphasis on sustainable design," in *Functional Reverse Engineering of Strategic and Non-Strategic Machine Tools*, 1st ed., W. Khan, K. Ghulam and H. Ghulam Abbas, Ed. Boca Raton: CRC Press, 2021.

[27] G. Ngaile and T. Altan, "Simulations of manufacturing processes: Past, present and future," *Proceedings of the Seventh ICTP*, October 2002, Japan, p 271.

[28] T. Howson and H. Delgado, "Computer modeling metal flow in forging," *JOM*, vol. 41, no. 2, pp. 32–34, 1989.

[29] S. Oh, "Finite element analysis of metal forming processes with arbitrarily shaped dies," *International Journal of Mechanical Sciences*, vol. 24, no. 8, pp. 479–493, 1982.

[30] C. Sellars, "Modelling microstructural development during hot rolling," *Materials Science and Technology*, vol. 6, no. 11, pp. 1072–1081, 1990.

[31] G. Shen, "Microstructure modeling of forged components of ingot metallurgy nickel based superalloys," in *Advanced Technologies for Superalloy Affordability*, K. Chang, S. Srivastava, D. Furrer and K. Bain, Ed. Warrendale, PA: TMS, 2021, pp. 223–231.

[32] O. Brucelle and G. Bernhart, "Methodology for service life increase of hot forging tools," *Journal of Materials Processing Technology*, vol. 87, no. 1–3, pp. 237–246, 1999.

[33] "Simufact material database," *Simufact*, 2021 [Online]. Available: www.simufact.com/material-data.html [Accessed 3 September 2021].

[34] "Material center databanks," *Mscsoftware.com*, 2021 [Online]. Available: www.msc-software.com/product/materialcenter-databanks [Accessed 3 September 2021].

[35] C. Chang and A. Bramley, "Determination of the heat transfer coefficient at the workpiece—die interface for the forging process," *Proceedings of the Institution of Mechanical Engineers, Part B: Journal of Engineering Manufacture*, vol. 216, no. 8, pp. 1179–1186, 2002.

[36] Catalogue. "SIJ Metal Ravne d.o.o.," 2021 [Online]. Available: https://sij.metalravne.com/steelselector/selector.html.

11 Milling Path Planning of Glasses Lens Edge Based on Dynamic Gaussian Smoothing

Hao Liu, Shuhao Xu and Zhengyin Chen

CONTENTS

11.1 INTRODUCTION

Due to myopia, presbyopia, eye protection, and beauty, glasses have become an indispensable item in people's daily life and an important factor affecting people's quality of life. The processing ability of the lens edge, including processing efficiency, quality and types, has an important impact on the quality of glasses. With the continuous expansion of digital manufacturing technology to various fields, the appearance of the lens digital edge milling system has greatly improved the edge processing capabilities of the lens, thereby being increasingly recognized, and valued by people. Opticians use such a system to cut and edge lens blanks. This enables the glasses to be "quickly prepared and then to be processed". It also can greatly improve the degree of fit between the lenses and the frame, increase the comfort of wearing the glasses, and extend the service life of the glasses.

DOI: 10.1201/9781003220985-11

The digital lens edger machine can perform various processes such as lens rough cutting, fine milling, sharp edge milling, slotting, chamfering, and milling step. It combines the functions of traditional manual lens cutting machines, ceramic wheel edging machines, and the drilling machine [1]. Currently, the universal digital edger machine in the international market is Mei system in German [2]. The basic processing motion mode of the system is the four-axis milling motion principle: AC turntable that can move up and down, left, and right, and high-speed rotating tool with horizontal fixed axis. The rotation of the C axis and the up-and-down movement of the turntable in the Z direction are the main movements, which can be understood as the polar angle and the polar radius of polar coordinates. Additionally, the pitch movement of the A axis and the left-and-right movement of the turntable along the X axis are auxiliary movements used to adjust the shape of the lens edge. Therefore, the data used by the lens edger machine system is usually polar coordinate data. The data transmission format complies with the OMA data standard established by ISO13666:2019 [3].

The initial data of the cutting line of the lens edge is usually derived from the digital scanning of the inner edge of the lens frame by the special scanning equipment of the lens frame [4,5]. Due to the impacts of the lens frame processing technology, welding, and spraying process, the polar radius in the OMA data file received by the CAM system usually contains much noise. Therefore, to ensure that the milling path is generated by the equidistant method for the cutter location, the polar coordinate data in the file must be de-noised. After de-noising, the data not only needs to be a smooth curve that meets the equidistant requirements, but also the polar angle distribution is as evenly as possible and the data is as high effectiveness as possible, which ensures the processing quality of the lens.

The de-noising of edge contour curve (also called smoothing) is a classic problem in the field of manufacturing and geometric design. The solutions can be divided into four main types: spline function fitting method, energy smoothing method, multi-resolution method, and direct smoothing method. For the first type of spline function fitting method, the basic principle is to use the spline curve method to fit the noise data and apply the Ck continuity of the spline curve to construct a smooth contour curve. For example, Calio et al. [6] used an integral spline with shape parameters to fit the curve to achieve an overall smoothing effect, and then adjusted the shape parameters locally in sections to achieve a balance between smoothing and approximation accuracy. Moreover, Hashemian et al. [7] used the NURBS method to fit the noise curve, and then used the curvature variational principle for optimizing the position of the control vertex. The method of Birk et al. [8] is like that of Hashemian et al. [7], except that they used B-splines and deduced a more concise calculation method from the perspective of curvature variation. Yang [9] achieved the balance between the smoothing effect and the fitting accuracy by adjusting the weight factors of the NURBS curve. Wang et al. [10] directly construct the control vertices of the B-spline curve in the optimization model through the curve fairing criterion. The idea is simpler than the previous methods. Wu et al. [11] combined the particle swarm algorithm with the BP neural network method and provided a method to solve the B-spline control vertex to achieve the smoothing effect. Liu et al. [12] designed a smoothing B-spline curve fitting algorithm only based on particle swarm optimization. For the second type of method, Yong et al. [13] provided a method to construct a smooth Ferguson curved portion based on the principle of minimum deformation energy. Zhang et al. [14] proposed a smoothing method of parametric cubic spline curve based on the principle of minimum deformation energy. Zhang [15] proposed an automatic smoothing algorithm for cubic B-spline curve based on local energy. For the third type of method, Amati [16], Ji et al. [17], Li et al. [18], and Yu et al. [19] successively proposed the smoothing methods of the curve based on the B-spline wavelet. For the fourth type of method, the Tabin method is a classic and effective method. It considered the defects of the curve contraction caused by the ordinary smoothing process and gave a non-shrinking smoothing algorithm. Liu et al. [20] constructed curvature scale space (CSS) based on the Gauss smoothing method to obtain smoothing curves at different scales. Moreover, Liu et al. [21] provided a method to directly adjust the position of discrete points based on the curvature calculation formula of the discrete curve. This method is more flexible. It can usually be combined with curvature variation method, energy method, etc., to achieve the purpose of smoothing the curve by iteratively solving the optimization model.

Although the above methods can achieve the purpose of removing curve noise, they are usually designed only based on the principles of graphics without the consideration of the needs of actual workpiece processing occasions. Compared with the above-mentioned literatures, Lin et al. [22] considered the problems of non-self-intersection and equidistance, which is very close to the working conditions of this paper. Li et al. [23] considered the fluency of the tool movement when smoothing the curve, which is also the problem that this paper needs to consider when planning the tool path.

Different from the above-mentioned literatures that only consider the curve smoothing problems based on the traditional fairing criterion [24,25], this paper considers the smoothing problem according to the characteristics of polar coordinate machine tool processing. In fact, not only the lens processing involved in this article requires a polar coordinate machine tool, but also our country has more applications of polar coordinate machine tools in other occasions. For instance, Bai et al. [26] conducted research on the control system of polar coordinate machine tools. Cao et al. [27] discussed the transmission system of the polar coordinate machine tool. Additionally, Zhou [28] and Chen [29] studied the CNC programming technology of the polar coordinate machine tool. Therefore, on the basis of summarizing the existing domestic and overseas curvilinear smoothing technology, and taking the lens processing conditions as the background, this paper designs a Gauss smooth curve dynamic smoothing method for the contour milling of polar coordinate machine tools. This method uses minimum smoothing steps to eliminate the noise of the edge curve in the lens frame. Based on keeping the original data within a given accuracy range, the polar angle of each point on the processing path is distributed as evenly as possible, so that the feed movement of the tool is smooth, thereby achieving the purpose of improving the edge quality of the lens workpiece.

11.2   BACKGROUND KNOWLEDGE

11.2.1   MACHINE TOOL STRUCTURE AND MOTION MODES

As shown in Figure 11.1(a), the glass-type four-axis milling machine includes a spindle for clamping tools. The end of the spindle is fixed on a sliding table that can slide left and right (that is, moves in the X direction). The spindle can rotate around the Y axis with its end as the center. For the machine tool developed in this paper, this axis of rotation is called the V axis. The lens is clamped on the C axis parallel to the X axis. This is like a lathe, and the C axis carries the lens to rotate linearly around its center. The device equipped with the C axis can move up and down, namely, moving along the Z axis. Figure 11.1(b) shows the coordinate system of the machine tool in the CAM software and displays initial relative positions of the lens and the tool. The tool has rotated an angle along the V axis.

11.2.2   OMA DATA FILE

Table 11.1 shows the basic structure of a typical OMA data file. The data in the plane data field is the basic data. Its meaning is to regard the lens center as the polar coordinate center. The units of the polar radius and polar angle of the lens edge are 0.01mm and 0.01 degree, respectively. Furthermore, Z and ZA represent the distance from the discrete points on the center line of the lens edge to the polar coordinate plane. Z0 and Z0A, Z1 and Z1A respectively denote the distances from the discrete points on the edge curve of the A surface and the edge curve of the B surface to the polar coordinate plane. When the data in OMA is converted to the machine tool coordinate system, Z, X, Y and X, Y, Z are fully symmetrically rotated.

11.2.3   PROCESSES AND TOOLS

Table 11.2 shows the commonly used processes for lens edge processing, that is, the corresponding tools. It can be seen from Table 11.2 that the outer equidistance of the closed contour on the polar coordinate plane formed by the polar radius and the polar angle is the key to the planning of the trajectory of the lens edge processing. This is because the machining command describes the trajectory of the tool location point (that is, the center of the tool cross section).

11.3   CONTOUR EQUIDISTANCE AND DYNAMIC SMOOTHING METHOD

11.3.1   CONTOUR EQUIDISTANCE

When calculating the equidistance of the tool location points for the three-dimensional profile represented by the OMA file, it is firstly assumed that the profile curve is a plane curve with Z = 0. The equidistance value d for the plane curve is determined according to the tool radius and machining allowance. After equidistant to the plane curve represented by polar coordinates, the Z coordinate value in the OMA file is added to the equidistant curve, thereby obtaining the three-dimensional equidistant

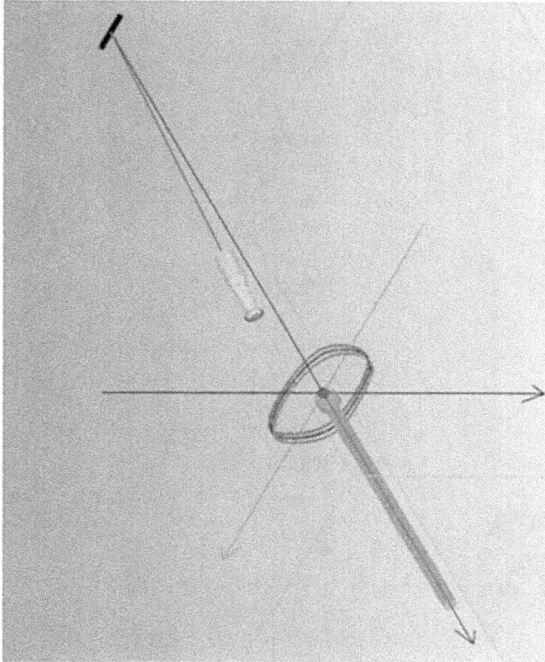

(b) Initial Relative Position of Lens and Tool in Machine Tool Coordinate System.

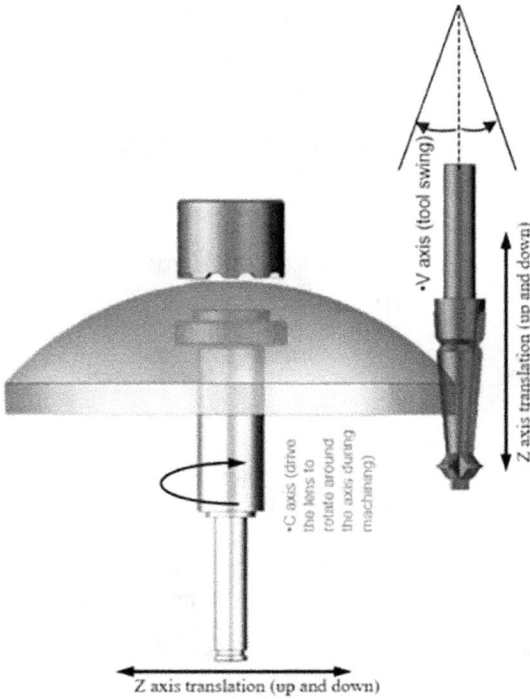

(a) Four-Axis Milling Machine for Glasses.

FIGURE 11.1 Four-Axis Milling Machine for Glasses and the Machine Tool Coordinate System.

TABLE 11.1

Basic Structure of OMA Data File.

Start region		LDCAM_FILE=jingpian1
		LDCAM_Unit_ID = B02
		LDCAM_Editor = wukai
		LDCAM_Version = 2.0
		LDCAM_Date = 20210617
		TRCFMT = 1;800;U;R;F
Informational region	Plane Data Field	R = 2994;2992;2991;2989;2988;2986;2985;2983
		. . . . . .
		R = 3008;3006;3004;3003;3001;2999;2997;2996
		A = 38;81;124;168;211;254;298;341
		. . . . . .
		A = 35693;35736;35779;35822;35865;35909;35952;35995
	Additional Data Field	ZFMT = 1;800;U;R;F
		Z = 182;181;180;180;179;178;177;176
		. . . . . .
		Z = 189;188;187;187;186;185;184;183
		ZA = 38;81;124;168;211;254;298;341
		. . . . . .
		ZA = 35693;35736;35779;35822;35865;35909;35952;35995
		Z0FMT = 1;800;U;R;F
		Z0 = 227;227;226;226;226;226;225;225
		. . . . . .
		Z0 = 229;229;228;228;228;228;227;227
		Z0A = 38;81;124;168;211;254;298;341
		. . . . . .
		Z0A = 35693;35736;35779;35822;35865;35909;35952;35995
		Z1FMT = 1;800;U;R;F
		Z1 = 634;634;633;633;632;632;631;630
		. . . . . .
		Z1 = 639;638;638;637;637;636;635;635
		Z1A = 38;81;124;168;211;254;298;341
		. . . . . .
		Z1A = 35693;35736;35779;35822;35865;35909;35952;35995

curve of the spectacles frame. Therefore, the equidistance of the plane curve is the core work of calculating the edge processing path of the glasses.

As shown in Figure 11.2, after converting the polar coordinate point to the point in the rectangular coordinate, this paper applies the exterior angle bisector method to calculate the normal vector of the curve. The movement distance of pi along the respective normal vector direction is

$$s = d / |\cos\theta|$$

TABLE 11.2

Lens Process Types and Processing Methods.

Cutting process type	Cutting tool and edge point	Swinging Angle of the cutting tool	Radius of the cutting tool (mm)	3d model display
Rough cut	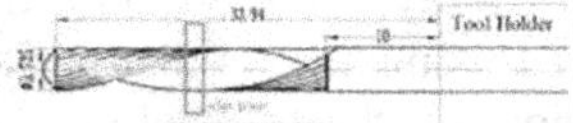 Tool Holder Machine Tool T01	V = 0	1.8675	
Fine cut	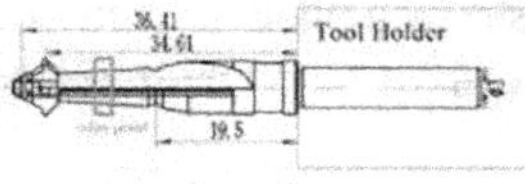 Tool Holder Machine Tool T02	V = 4	2.75	
groove	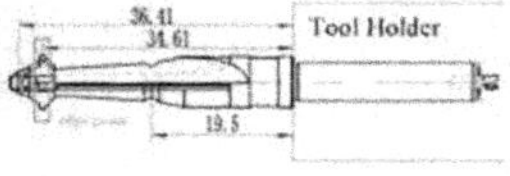 Tool Holder Machine Tool T02	V = 0	3.3125	
step	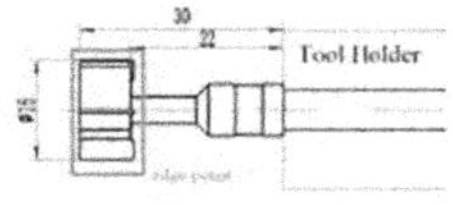 Tool Holder Machine Tool T03	V = 0	7.5	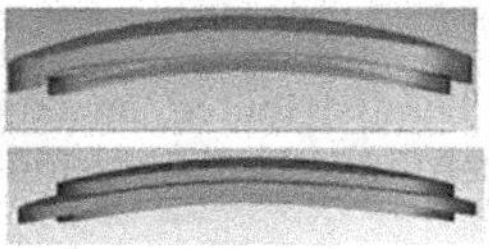
chamfering	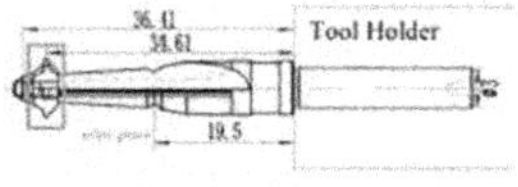 Tool Holder Machine Tool T02	V = 40	2.7	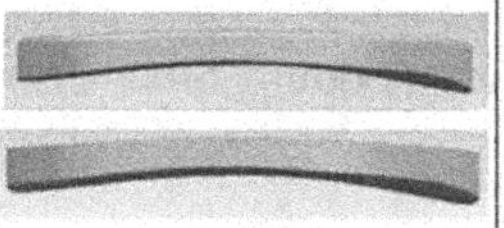
sharp edge	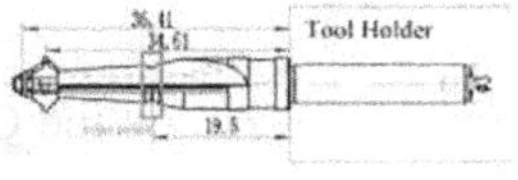 Tool Holder Machine Tool T02	V = 4	3.15	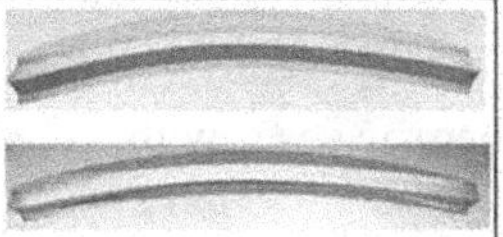

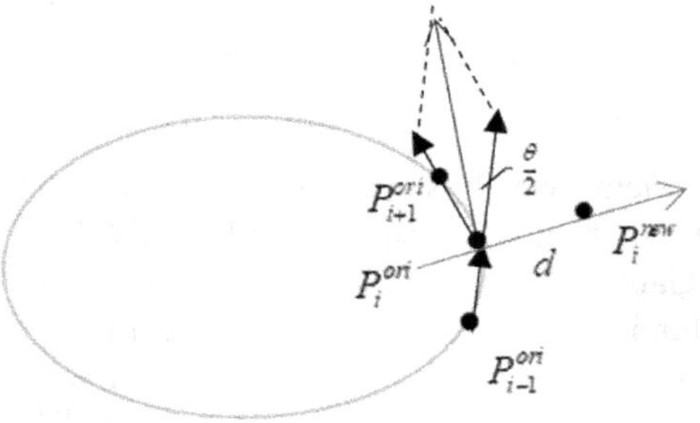

FIGURE 11.2 Schematic Diagram of the Normal Vector of the Curve Calculated by the Exterior Angle Bisector Method.

The advantage of this method is that the equidistance is intuitive and accurate, and it can be close to zero error for noise-free data. It is especially suitable for the outer equidistance of the convex curve represented by the lens profile. Because this equidistance does not generate self-intersection, there is no need to consider the clipping problem caused by self-intersection. Even if the self-intersection of the equidistant curve occurs in the inner equidistant, it can be easily solved through the calculation of the intersection point of the normal rays on the equidistant side.

On the other hand, the disadvantage of this method is that the accuracy of the equidistant curve is very sensitive to the noise of the original curve. For the curve represented by polar coordinates, there is usually a phenomenon of self-intersection. That is, assume that $p_i^{ori}(\rho_i^{ori},\theta_i^{ori})$ and $p_{i+1}^{ori}(\rho_{i+1}^{ori},\theta_{i+1}^{ori})$ are two adjacent points on the OMA data file, $p_i^{off}(\rho_i^{off},\theta_i^{off})$ and $p_{i+1}^{off}(\rho_{i+1}^{off},\theta_{i+1}^{off})$. If

$$\theta_i^{off} \geq \theta_{i+1}^{off}$$

Then, the equidistant curves form self-intersections.

Generally, in order to ensure the smooth rotation of the C axis of the machine tool and the smooth edge of the lens processing, not only the dispersion curve $p_i^{off}(\rho_i^{off},\theta_i^{off})$ does not form self-intersection, but also $\theta_i^{off}(i=0,1,\cdots,N)$ is required to be distributed as equally spaced as the $\theta_i^{on}(i=0,1,\cdots,N)$ in the OMA file. Therefore, the dispersion curve $p_i^{ori}(i=0,1,\cdots,N)$ needs to be smoothed.

11.3.2 Non-Shrinking Gaussian Smoothing Method

The smoothing of the dispersion curve $p_i^{on}(i=0,1,\cdots,N)$ is usually a weighted average:

$$q_i = \sum_{j=-k}^{k} w_j p_{i+j}$$

where $w_j \geq 0$, with $\sum_{j=-k}^{k} w_j = 1$, $p_{-5}=p_{N-5}$ and $p_{N+5}=p_5$. As the Guass smoothing operator has fine scale-dependency and its defined weights perfectly match the normal distribution, this paper uses the Guass smoothing function to define the weights $w_j(i=0,1,\cdots,N)$:

$$G(x) = \frac{1}{\sqrt{2\pi}\sigma} e^{-\frac{x^2}{2\sigma^2}}$$

where σ is the scale factor, which is generally taken as 0.8~2. The larger the scale factor, the larger the k in the discrete smoothing formula (1).

Considering that Gaussian smoothness can cause the curve to shrink, the smoothing principle of Taubin is used for reference. This paper presents a curve smoothing process without shrinkage. Namely, assume that (i) W is the Gaussian smooth operator matrix defined by equation (11.1) for the curve; (ii) K=IW; (iii) P is the column vector formed by all points on the discrete curve; and (iv) and Q is a sequence of points after smoothing without shrinkage. Then, a smoothing step without shrinkage can be expressed as:

$$\begin{cases} Q = (I - \mu K)(I - \lambda K)P \\ \dfrac{1}{\lambda} + \dfrac{1}{\mu} = k_{PB} \approx 0.1 \\ \lambda > 0 \end{cases} \qquad (11.3)$$

In the above smoothing step, λ is determined empirically when the above conditions are met, and μ is calculated according to the value of λ using the above formula. In fact, the value of λ is related to the smoothing scale σ in formula (2). When σ takes a larger value, λ takes a smaller value. Otherwise, because the moving distance of each discrete point in a smoothing step is relatively large, irregular folds can appear in the curve after a few smoothing steps. In this paper, $\lambda = \sigma = 1.0$ is selected. The μ obtained by the above formula is usually greater than 1. Furthermore, when the distance moved along the normal direction or the direction of the Laplace operator is large, the possibility of local deformity of the curve is relatively large. Hence, the method in this paper is further improved based on the Taubin method:

$$\mu = c / (k_{PB} - 1 / \lambda)$$

where the constant coefficient c is suitable to be 0.5~1.0. In the test of this paper, it is set to 0.75, so that the moving distance along the direction of the Laplace operator is small to ensure the stability of iteration. In the following discussion, the non-shrinking smoothing operator defined by equation (11.3) is recorded as $TaubinSmooth(\bullet)$.

11.3.3 DYNAMIC SMOOTHING METHOD

Even if the non-shrinking smoothing operator $TaubinSmooth(\bullet)$ is used, it can still smooth out the detailed features in the contour, and gradually make the curvature of each point on the contour tend to be the same. Thus, the smoothing error between the smoothing curve and the original curve is bound to exist. At the same time, to make the polar angles of discrete points on the equidistant curve as evenly distributed as possible, the curve needs to be as smooth as possible. The balance condition of the two is to make the smoothing times of the edge curve of the lens as small as possible when the polar angle distribution meets the conditions. For this reason, this paper designs a dynamic smoothing method:

$$\min t$$
$$s.t. \ q_i^0 = p_i^{ori}$$
$$p_i^{off} = offset(q_i^t)$$
$$\delta_{max} \geq \theta_{i+1}^{off} - \theta_i^{off} \geq \delta_{min} > 0$$
$$HausdorffDis(Q^0, Q^i) \leq \varepsilon$$
$$q_i^t = TaubinSmooth(q_i^{t-1})$$
$$i = 0, \cdots, n-1$$
$$t = 0, 1, 2, 3, \ldots.$$

Where t is the number of smoothing, and $offset(\bullet)$ represents the equidistant operator defined in Figure 11.2. δ_{max} and δ_{min} is the maximum and minimum allowable polar angle spacing. These are two empirical values, which are determined by the performance of the edge cutter. $\delta_{min} > 0$ is required, when cutting the lens, the C axis cannot be rotated, otherwise there is a high possibility that the edge will not be smooth. ε denotes the assembly tolerance between the lens and the frame, which is determined by the assembly process. $HausdorffDis(\bullet,\bullet)$ represents the shortest distance between two-point sets. In this paper, a loop iterative process is used to solve the above optimization model. Once the constraints are met, the iteration stops and the required dispersion curve is obtained.

11.4 EXPERIMENT AND DISCUSSION

To verify the correctness of the algorithm in this paper, a self-developed four-axis polar coordinate lens digital edge cutting machine is applied for the verification of the relevant data, as shown in Figure 11.3(a). The speed of the tool in the machine tool is r/s. The edge of the lens is cut by a micro-combined tool or an end mill to ensure that the edge is smooth with low noise in the edge data. The C axis speed is dynamically adjusted according to the distance from the tool contact point to the center of rotation. The feed rate of the tool relative to the workpiece is set to be constant. The processing path adopts linear interpolation. The lens edge defects will be obvious when the path is noisy or the acceleration and deceleration of tool feed are uneven. For this equipment, select $\varepsilon = 0.01mm$, $\delta_{max} = 0.8$, and $\delta_{min} = 0.2$. Because the control system of the equipment adopts linear interpolation, a larger δ_{max} can lead to uneven edge contours, and a smaller δ_{min} can cause discontinuous rotation of the C axis due to frequent acceleration and deceleration. Matlab2014a is used for numerical analysis. The computer configuration used for calculation time analysis is Intel(R)Core(TM)i5–1400F CPU processor and 64-bit operating system.

(a) Equipment appearance (b) Basic structure of the equipment

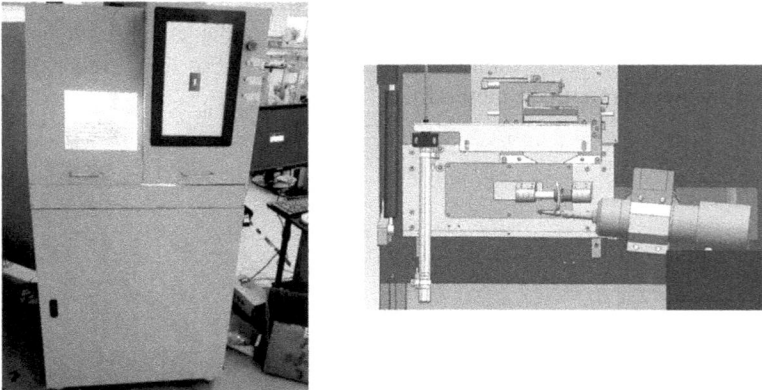

FIGURE 11.3   Four-Axis Polar Coordinate Lens Digital Edge Cutting Machine.

11.4.1 ANALYSIS AND COMPARISON OF EXTERIOR ANGLE BISECTOR EQUIDISTANT METHOD

In order to verify the accuracy of the exterior angle bisector equidistance method in this paper, this paper takes 800 equi-polar angle distribution points on the ellipse circumference and the spectacle lens profile designed by UG and uses the method in this paper to make equidistance lines. Subsequently, the original contour and the equidistant curves generated by this method are imported into AutoCAD software. The isometric error of the method is analyzed by AutoCAD software. Figure 11.4 shows the isometric effect of the exterior angle bisector equidistant method on the noise-free ellipse data and the AutoCAD data of the noise-free lens.

As shown in Figure 11.4, it can be found that the exterior angle bisector equidistant method is robust and stable to noise-free contour data with uniform polar angle distribution. The error is within 0.01mm, which fully meets the processing requirements of the lens edge. At the same time, because the data collected by the spectacles frame scanning device in the application is noisy, to make this equidistant algorithm work stably, it is necessary to apply a smoothing algorithm to filter out the scanning noise.

In the algorithm for calculating the normal vector of the discrete curve, the parameter cubic spline method is also a commonly used method. Moreover, the normal vector is associated with the position information of all points by solving the equation group when calculating the normal vector. Figure 11.5(a) shows the effect comparison of the noise-free ellipse, the radius-length random noise curve, and the scan curve with small noise when using two algorithms to calculate the normal vector. Figure 11.5(b) indicates that the calculation results of the two methods have a deviation of 0.3 degrees. According to formula (1), the magnification factor of equidistant length is

$$c = 1/\cos(0.3\pi/180/2) = 1.000003426955759$$

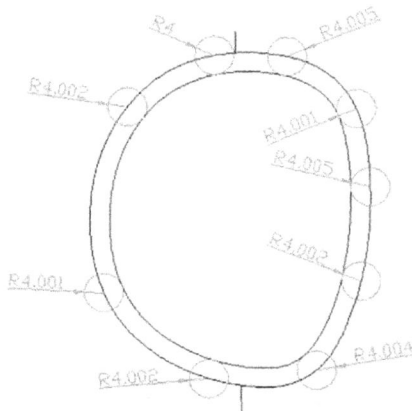

FIGURE 11.4 Isometric Effect of AutoCAD Data of Noise-Free Lens.

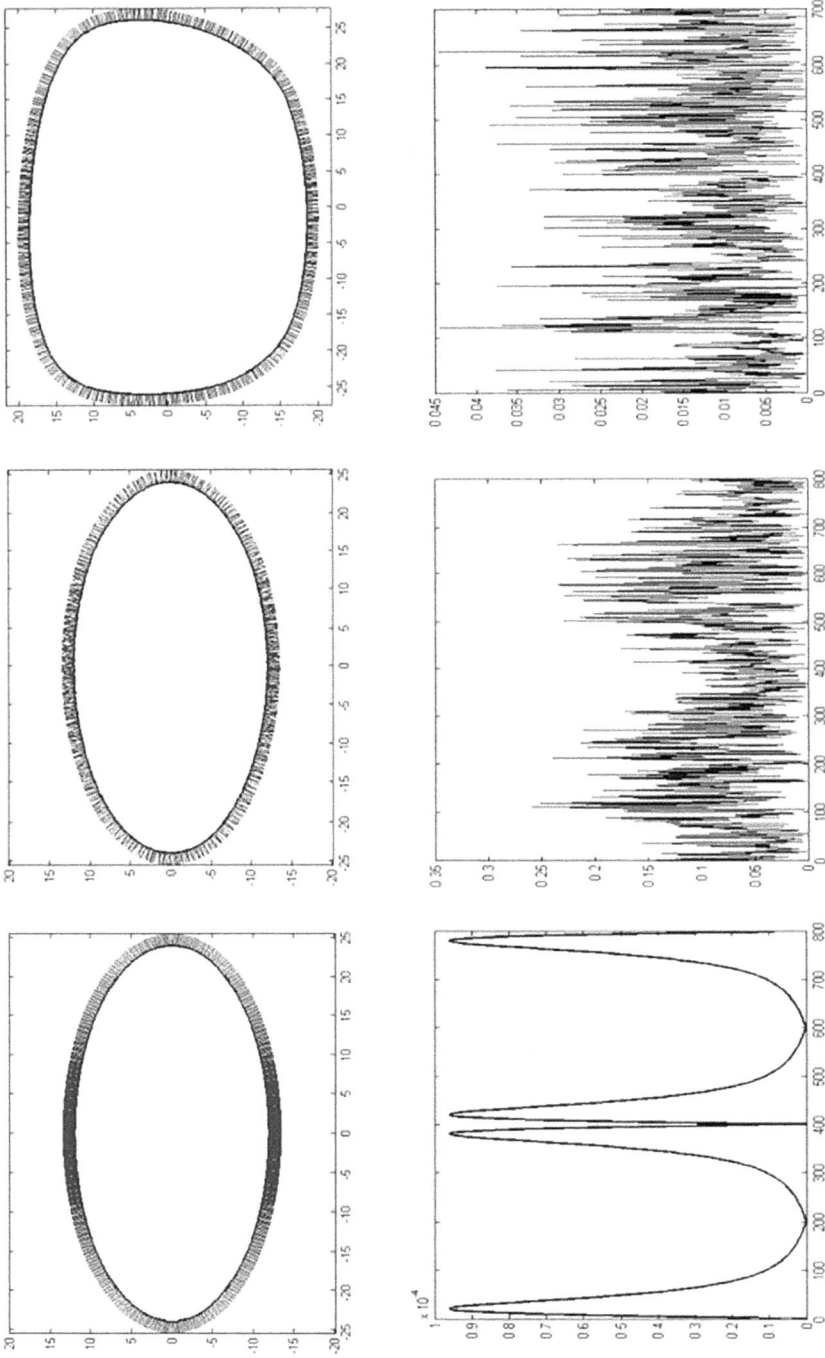

(a) Noise-free ellipse, diameter-length random noise curve, scanning small noise curve and its normal vector

(b) Angle between normal vectors of two adjacent points (forward-difference comparison)

FIGURE 11.5   Evaluation the algorithm using three different curves.

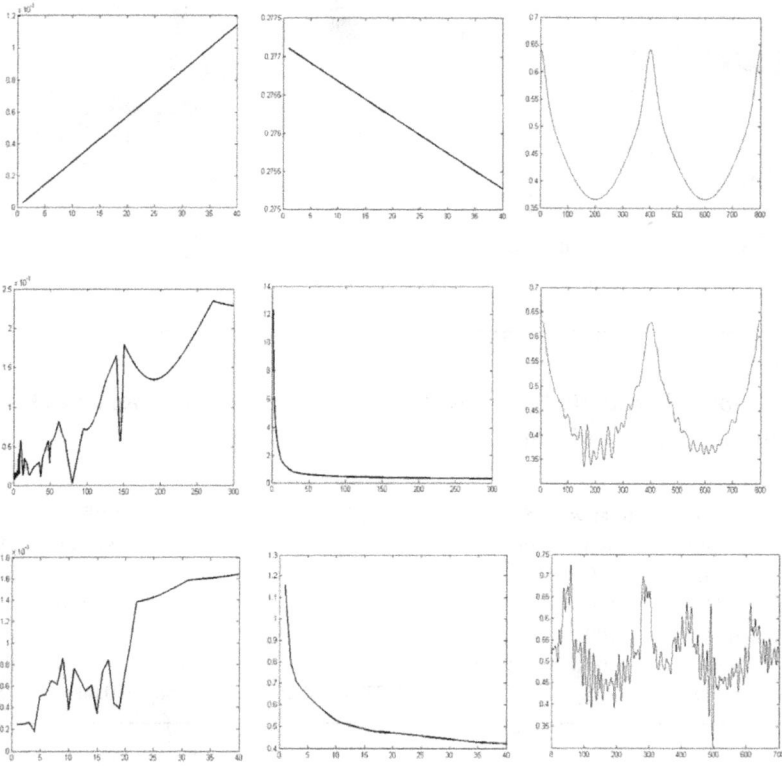

(a) Noiseless data (40 iterations)
(b) Artificially added noise data (300 iterations)
(c) Scanned small noise data (40 iterations)

FIGURE 11.6 Influence of iteration times of smoothing algorithm on convergence of smoothing approximation error and corner distribution in noiseless data, artificially added noise data and scanned small noise data (left: smooth errors; middle: maximum angle; right: angle distribution for a iteration step).

This demonstrates that there is no difference between the two within the scope of application. However, the parametric cubic spline method needs to solve the tridiagonal linear equations. This means that the consumption on calculation memory and time is larger than that of the exterior angle bisector method.

11.4.2 ITERATIVE CONVERGENCE ANALYSIS

This paper takes three groups of data, including designed noise-free data, artificially added noise data, and scanned small noise data, and observes the convergence of the iteration for the fairing approximation error as well as the distribution of the rotating angle. Figure 11.6 indicates that the dynamic iterative smoothing method

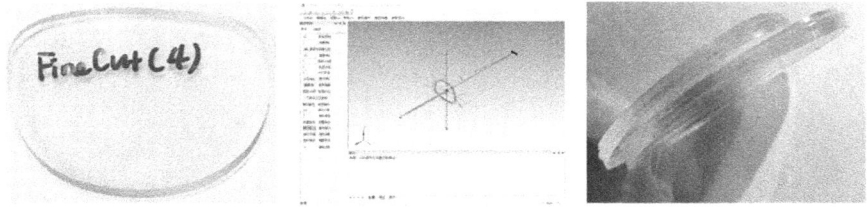

FIGURE 11.7 Physical Map of L\ens.

TABLE 11.3
The Number of Discrete Points, Iteration Steps, Iteration Time, and Other Data of the Five Sets of Lens Data.

| | Discrete point | Iterative step | Smoothing error (mm) | Adjacent angle (°) | | Time (s) | | |
				Max	Min	Error	Other	Total
OMA1	700	100	0.0016	1.3404	0.3081	924.6574	0.0583	924.7156
OMA2	720	100	0.0019	0.8950	0.3626	961.6837	0.0570	961.7407
OMA3	720	100	0.0024	0.9161	0.3399	999.4323	0.0584	999.4907
OMA4	720	69	0.0021	0.7995	0.3203	663.1121	0.0400	663.1521
OMA5	700	5	0.0003	0.7274	0.2593	45.1589	0.0063	45.1652

in this paper has good calculation stability. This method can also obtain satisfactory calculation results for any type of data and obtain a balance between the fairing error and the distribution of the rotating angle. In addition, for noise data, the error diffusion speed is slow, and the differential of the rotating angle distribution converges. Especially for the scan data of the lens frame edge, the number of iterations is less than that of the artificial noise data, and the rotating angle distribution meets the needs of lens edge cutting. This means that the algorithm design in this paper is reasonable for the edge cutting of the lens.

11.4.3 Smoothing Step and Iteration Time Analysis

With five sets of data taken, the scan data is collected by the spectacles frame equipment. The distribution of the smoothing error and the rotating angle difference, number of discrete points, iteration steps, iteration time, and other data are shown in Table 11.3. The designed stopping criterion is that the rotating angle difference is distributed within 0.25~0.8, or the number of iterations is greater than 100, or the fairing error is greater than the given value. It can be seen from Table 11.3, that for general scan data, the requirements can be met within 100 smoothing steps. For data 1~3, at the maximum iteration step, the maximum difference between adjacent rotating angle is still greater than 0.8 degrees. This is because the area near the highest point of the frame edge is usually small in polar radius and close to a straight line. The fairing error is within the given range, and the equidistant curve has no self-intersection (this is because the smallest rotating angle difference is greater than 0.2, and no rotation phenomenon occurs on the C axis). Therefore, the smoothing data can be post-processed, and the line segment corresponding to the large polar angle can be divided into several shorter line segments, so that the polar angle distribution of the edge profile meets the requirements. The calculation of Hausdorff distance is the bottleneck of the iteration time. The reason is that this paper adopts the algorithm of calculating the distances between points and line segments for the calculation of the Hausdorff distance. The time complexity is $O(n^2)$, where n is the number of points in the discrete curve. Since the fairing error within 100 smoothing steps must be within the assembly tolerance, the calculation of the Hausdorff distance can be abandoned in actual application to greatly improve the calculation efficiency.

REFERENCES

[1] Shanqi, Zeng, Nan, Guo and Zhenya, Tian, 2009. New NC lens cutting machine design. *Machinery Design & Manufacture*, 4, pp. 29–30.

[2] Mei, Srl. *MEI System: Homepage[Z/OL]*. https://meisystem.com.

[3] Comité Européen de Normalisation. *Ophthalmic optics—Spectacle lenses—Vocabulary: EN ISO 13666–2019 [S]*. www.iso.org.

[4] Xu, Feng, 2005. *Numerical control lens-edging system hardware design and tracer software Development*. Nanjing University of Aeronautics and Astronautics. www.nuaa.edu.cn.

[5] Guo, Caiping, 2009. *Edge Detection System of Lens Based on Linear Array CCD*. North University of China.

[6] Caliò, F., Miglio, E. and Rasella, M., 2010. Curve fairing using integral spline operators. *International Journal for Numerical Methods in Biomedical Engineering*, 26(12), pp. 1674–1686.

[7] Hashemian, A. and Hosseini, S.F., 2018. An integrated fitting and fairing approach for object reconstruction using smooth NURBS curves and surfaces. *Computers & Mathematics with Applications*, 76(7), pp. 1555–1575.

[8] Birk, L. and McCulloch, T.L., 2018. Robust generation of constrained B-spline curves based on automatic differentiation and fairness optimization. *Computer Aided Geometric Design*, 59, pp. 49–67.

[9] Yang, X., 2018. Fitting and fairing Hermite-type data by matrix weighted NURBS curves. *Computer-Aided Design*, 102, pp. 22–32.

[10] Shiwei, Wang, Ligang, Liu Juyong, Zhang, et al., 2016. Sparity optimization based curve fairing. *Journal of Computer-Aided Design and Computer Graphics*, 28(12), pp. 2043–2051.

[11] Yize, Wu, Xu, Zhang, Mingyang, Jiang, et al., 2018. Curve smoothing algorithm for particle swarm optimization BP neural network. *Control and Instruments in Chemical Industry*, 45(12), pp. 939–942, 954.

[12] Wufei, L.I.U. and Xu, Z.H.A.N.G., 2020. Fairing reconstruction algorithm of B-spline curve based on PSO. *Journal of Light Industry*, 35(2).

[13] Yong, J.H. and Cheng, F.F., 2004. Geometric Hermite curves with minimum strain energy. *Computer Aided Geometric Design*, 21(3), pp. 281–301.

[14] Zhang, C., Zhang, P. and Cheng, F.F., 2001. Fairing spline curves and surfaces by minimizing energy. *Computer-Aided Design*, 33(13), pp. 913–923.

[15] Hudong, Zhang, 2016. An automatic faring algorithm of cubic B-spline curves based on local energy. *Journal of Xi' an Aeronautical University*, *34*(1), pp. 79–81, 85.

[16] Amati, G., 2007. A multi-level filtering approach for fairing planar cubic B-spline curves. *Computer Aided Geometric Design*, *24*(1), pp. 53–66.

[17] Ji, X.G., 2020. Rapid wavelet construction of multi-resolution fairing for curves and surfaces with any number of control vertices. *Advances in Mechanical Engineering*, *12*(6), p. 1687814020936382.

[18] Li, A.M. and Tian, H.B., 2012. A multiresolution fairing approach for NURBS curves. In *Applied Mechanics and Materials* (Vol. 215, pp. 1205–1208). Trans Tech Publications Ltd.

[19] Taubin, G., 1995, September. A signal processing approach to fair surface design. In *Proceedings of the 22nd Annual Conference on Computer Graphics and Interactive Techniques* (pp. 351–358). ACM. https://dl.acm.org.

[20] Liu, H., Dai, N., Zhong, B., Li, T. and Wang, J., 2017. Extract feature curves on noisy triangular meshes. *Graphical Models*, *93*, pp. 1–13.

[21] Liu, G.H., Wong, Y.S., Zhang, Y.F. and Loh, H.T., 2002. Adaptive fairing of digitized point data with discrete curvature. *Computer-Aided Design*, *34*(4), pp. 309–320.

[22] Xujun, Lin, Shuyou, Zhang, Jin, Wang, et al., 2019. Generating method of non-uniform rational B-splines equidistance curves with self-intersection and adjustable smoothness. *Computer Integrated Manufacturing Systems*, *25*(8), pp. 1920–1926.

[23] Wensen, Li, Shengqi, Guan, Lu, Zheng, et al., 2019. A smoothing method for tool path with G2 continuity based on clothoid curves. *Journal of Xi' an Polytechnic University*, *33*(1), pp. 88–94.

[24] Yong, Chen, Xu, Zhang, Yize, Wu, et al., 2019. Research on curve fairness of body panels. *Computer Era*, *4*, pp. 5–8.

[25] Chen, Yan, Xiaogang, Ji, Yichao, Yu, et al., 2019. Fairness factor method for quantitative analysis of curve fairness. *Mechanical Science and Technology for Aerospace Engineering*, *38*(5), pp. 736–741.

[26] Dapeng, Bai, Yukun, Li and Nan, Li, 2008. Research on motion control of machine tools with polar coordinate system. *Journal of Yanshan University*, *32*(3), pp. 206–208.

[27] Lihong, Cao, Gang, Li and Hong, Wang, 2012. The design of feed drive system for polar NC machine. *Journal of Mechanical Transmission*, *36*(4), pp. 67–69.

[28] Yue, Zhou, 2014. Polar coordinate features in direct programming on CNC machine tool. *Coal Mine Machinery*, *35*(2), pp. 107–108.

[29] Weifeng, Chen, 2013. A polar coordinate interpolation instruction used skillfully to process special-shaped parts on lathe and milling compound machine tools. *Occupation*, *11*, pp. 135–136.

12 MATLAB® Simulation of Various Sliding Mode Control Techniques for Practical Applications

Ali Nasir

CONTENTS

12.1 INTRODUCTION

Sliding mode control has been around since the 1960s. It is unique in a sense that it considers the bounded uncertainty in the design of the controller. There have been many applications of the sliding mode control including aerospace, electric machinery, robotics, and industrial automation. A major issue involved with the classical sliding mode control design is the issue of chattering. Chattering is a phenomenon where the control signal oscillates at a very high frequency (but usually with a low magnitude of oscillation). To overcome the chattering and to improve the control system performance, many variations have been proposed in the sliding mode control [1]. In this chapter, we discuss three main variations in the sliding mode control, i.e., higher order sliding mode control [2], terminal sliding mode control [3], and super-twisting sliding mode control [4]. The chapter begins with basic concepts of the sliding mode control [5]. The main idea of the sliding mode control design is to convert the control objective (stabilization or reference tracking) into reaching a manifold within the state space (called the sliding surface) in which the dynamics of the system are favorable in achieving the control objective (or one may say that the dynamics of the system *slide* into achieving the control objective once the state variables have values within the sliding surface).

Major contribution of this chapter is to guide the reader regarding the simulation of various sliding mode controllers using MATLAB®. This has been accomplished

DOI: 10.1201/9781003220985-12

179

with the help of solved examples for which the MATLAB code has been provided in the appendix. The control laws and relevant dynamics of the system to be controlled (where applicable) are provided. The details of "why" the controller works have been left out but the interested readers may refer the references cited in various sections of this chapter.

The scope of this chapter does not encompass all of the variations of the sliding mode control. There are many types of sliding mode control that have not been covered in this chapter due to limited availability of space. For example, the switched higher order sliding mode control [6] is a recent advancement in the higher order sliding mode control that has been discussed in this chapter. Similarly, an adaptive version of the terminal sliding mode control is found in [7]. Note that the terminal sliding mode that we have included in this chapter is not adaptive. Usually, adaptation involves introduction of the dynamic gains within the control law in order to achieve best performance under changing conditions. A novel reaching law for sliding mode control application in controlling the permanent magnet synchronous machines is reported in [8]. This reaching law not only reduces the chattering in the control but is also effective in fast reaching of the state variable values to be within the sliding surface. A chattering free digital sliding mode control with associated state observer has been reported in [9]. Similarly, an output feedback integral optimal sliding mode controller for magnetic levitation systems has been proposed in [10]. Another variation in the terminal sliding mode control for magnetic bearing applications has been reported in [11]. Another interesting design of sliding mode controller based on backstepping for quadrotor applications has been discussed in [12].

Despite many of the sliding mode control approaches not being discussed in this chapter, the chapter still provides value for the reader in getting started on the conceptualization and MATLAB simulation of various sliding mode controller design schemes. Once the reader is familiarized with the control schemes and the corresponding MATLAB codes discussed in this chapter, self-learning can be employed in enhancing the knowledge bank and understanding the latest variations and improvements in the mainstream control design methods.

12.1.1 BASICS OF SLIDING MODE CONTROL

The distinguishing feature of sliding mode control is its ability to cater for unknown bounded modeling uncertainties and disturbances in the system. For example, consider the following system

$$\dot{x}_1 = x_2 \quad \dot{x}_2 = f(x_1, x_2) + d(t) + u \tag{12.1}$$

Where, $x = \begin{bmatrix} x_1, x_2 \end{bmatrix}^T$ is the state vector, $f(x_1, x_2)$ is the state dynamics function, $d(t)$ is the disturbance and u is the control input signal. Furthermore, suppose that the state dynamics function and the disturbance signal have known and unknown parts, i.e.,

$$f(x_1, x_2) = f_m(x_1, x_2) + \Delta f(x_1, x_2) \quad d(t) = d_m(t) + \Delta d(t) \tag{12.2}$$

Where $f_m(x_1, x_2)$ is the known part of $f(x_1, x_2)$ and $\Delta f(x_1, x_2)$ is the unknown part. Similarly, $d_m(t)$ is the known part of $d(t)$ and $\Delta d(t)$ is the unknown part. We assume that the unknown parts in the system are always bounded, i.e.,

$$|\Delta f(x_1, x_2)| \le \alpha(x_1, x_2) > 0, \ |\Delta d(t)| \le \beta(t) > 0, \forall t \qquad (12.3)$$

Here $\alpha(x_1, x_2)$ and $\beta(t)$ are known bounding functions, Now, given the system described above, our control objective is to achieve asymptotic reference tracking for a desirable value x_{1d} of the state variable x_1, i.e., we want $x_1 \to x_{1d}$ as $t \to \infty$. For this purpose, we define an error signal

$$\epsilon = x_1 - x_{1d} \qquad (12.4)$$

Now, based on the above error, we define the key ingredient of a sliding surface controller, i.e., the sliding surface

$$S = \dot{\epsilon} + \lambda \epsilon, \lambda > 0 \qquad (12.5)$$

Here λ is a positive constant. There are two intuitive ideas behind the above definition of the sliding surface. First, the sliding surface is defined in such a way that whenever $S = 0$, $\dot{\epsilon} = -\lambda \epsilon$ which renders $\epsilon \to 0$ as $t \to \infty$ that implies $x_1 \to x_{1d}$ as $t \to \infty$. Hence, the sliding surface is selected in such a way that for all the values of states that render $S = 0$, the control objective is achieved (which means that the control objective has been converted into a new objective that is to achieve $S \to 0$). Second intuitive idea in the design of the sliding surface is that it's first derivative involves the control input, i.e.,

$$\dot{S} = \ddot{\epsilon} + \lambda \dot{\epsilon} = \ddot{x}_1 - \ddot{x}_{1d} + \lambda(\dot{x}_1 - \dot{x}_{1d})$$

$$\to \dot{S} = f(x_1, x_2) + d(t) + u - \ddot{x}_{1d} + \lambda(x_2 - \dot{x}_{1d}) \qquad (12.6)$$

Notice that the control signal u appears in the derivative of the sliding surface. The advantage of this situation is that we can choose u in such a way that the dynamics of the sliding surface render $S \to 0$ as $t \to \infty$. Now, we design the control in two steps. In the first step, we cancel out all the known terms, i.e.,

$$u = -f_m(x_1, x_2) - d_m(t) + \ddot{x}_{1d} - \lambda(x_2 - \dot{x}_{1d}) + v \qquad (12.7)$$

In above equation, v is part of the control input u that is to be designed to cater for the unknown terms Δf and Δd. Designing the control as in (7), results in the following dynamics of the sliding surface

$$\dot{S} = \Delta f(x_1, x_2) + \Delta d(t) + v \qquad (12.8)$$

Now, to make sure that $S \to 0$, we need $S\dot{S} < 0$. For this purpose, the following observation is used (based on (3) and (8))

$$\dot{S} \le \alpha(x_1, x_2) + \beta(t) + v \tag{12.9}$$

Above equation implies that

$$S\dot{S} \le S(\alpha(x_1, x_2) + \beta(t) + v) \tag{12.10}$$

Now, we select v to ensure $S\dot{S} < 0$ as follows

$$v = -(\eta + \alpha(x_1, x_2) + \beta(t)) Sign(S) \tag{12.11}$$

Where,

$$Sign(S) := \begin{cases} 1 & S > 0 \\ -1 & S < 0 \\ 0 & S = 0 \end{cases}$$

Substituting (11) into (10), we get

$$S\dot{S} \le -\eta Sign(S) S < 0 \tag{12.12}$$

Hence the final control law that renders $x_1 \to x_{1d}$ as $t \to \infty$ is

$$\begin{aligned} u = &-f_m(x_1, x_2) - d_m(t) + \ddot{x}_{1d} - \lambda(x_2 - \dot{x}_{1d}) \\ &-(\eta + \alpha(x_1, x_2) + \beta(t)) Sign(S) \end{aligned} \tag{12.13}$$

Notice the dependence of above control law on the sliding surface S and the bounds on the uncertainties $\alpha(x_1, x_2)$ and $\beta(t)$. Now we look at an example of the sliding mode based on the above discussion

Example one

Consider the following system

$$\dot{x}_1 = x_2, \ \dot{x}_2 = x_1^2 + 2x_2 + u + 0.1e^{-t} + \Delta_1(x_1, x_2) + \Delta_2(t)$$

Given that the uncertainty in the system is bounded by the following

$$\Delta_1(x_1, x_2) \le |x_1 + x_2| \ \Delta_2(t) \le \frac{0.1}{t+1}$$

Design a sliding mode control law that renders $x_1(t) \to \sin(t)$

Solution

Comparing the given information with the discussion above, we note that $\alpha(x_1, x_2) = |x_1 + x_2|, \beta(t) = \frac{0.1}{t+1}, f_m(x_1, x_2) = x_1^2 + 2x_2, d_m(t) = 0.1e^{-t}, x_{1d} = \sin(t)$. Now, let's pick the sliding surface to be as in (5) with $\lambda = 1$, i.e.,

$$S = x_2 - \dot{x}_{1d} + x_1 - x_{1d} = x_1 + x_2 - \sin(t) - \cos(t)$$

Now, picking $\eta = 1$, the control law is designed as

$$u = -x_1^2 - 2x_2 - 0.1e^{-t} - \sin(t) + \cos(t) - x_2 - \left(1 + |x_1 + x_2| + \frac{0.1}{t+1}\right) Sign(S)$$

Now, for simulation of the above controller in MATLAB, we take $\Delta_1(x_1, x_2) = 0.5|x_1 + x_2|$ and $\Delta_2(t) = \frac{0.01}{t+1}$. The results are shown in Figures 12.1–12.3 for zero initial conditions. The reference is tracked asymptotically. The control signal shows the well-known chattering problem of the sliding mode control. Sliding surface graph indicates that the reaching phase takes about half a second.

MATLAB Code

The MATLAB code for the example is provided in the appendix:

Exercise

In example 1, vary the values of η and λ and study the effect of the changes in these values on the convergence of the sliding surface to zero. Also, vary the expressions used for $\Delta_1(x_1, x_2)$ and $\Delta_2(t)$ in the simulation (without violating the bounds) and study the effect on the reference tracking performance of the controller.

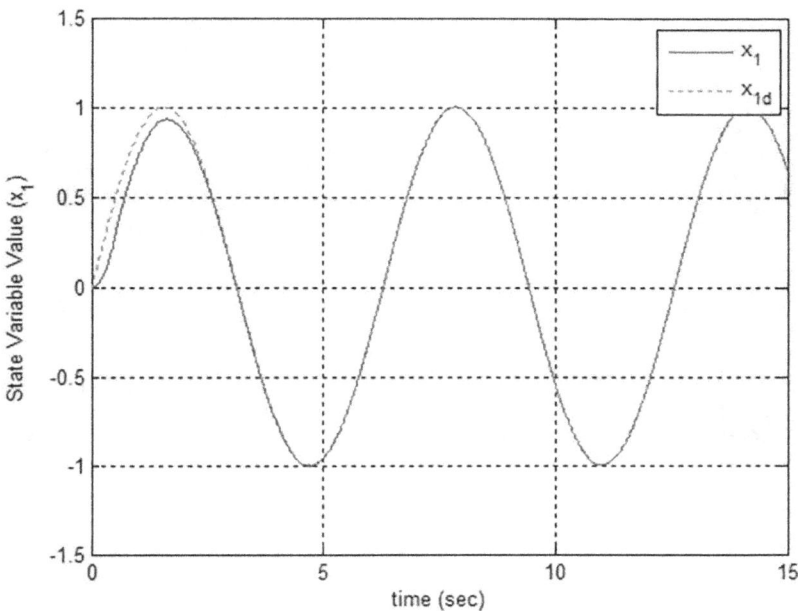

FIGURE 12.1   Reference Tracking in Example One.

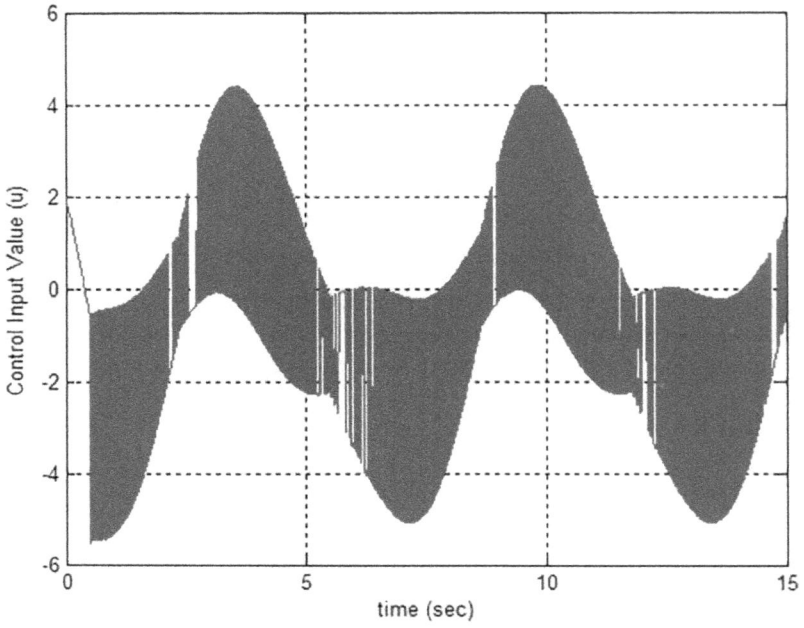

FIGURE 12.2   Control Signal in Example One.

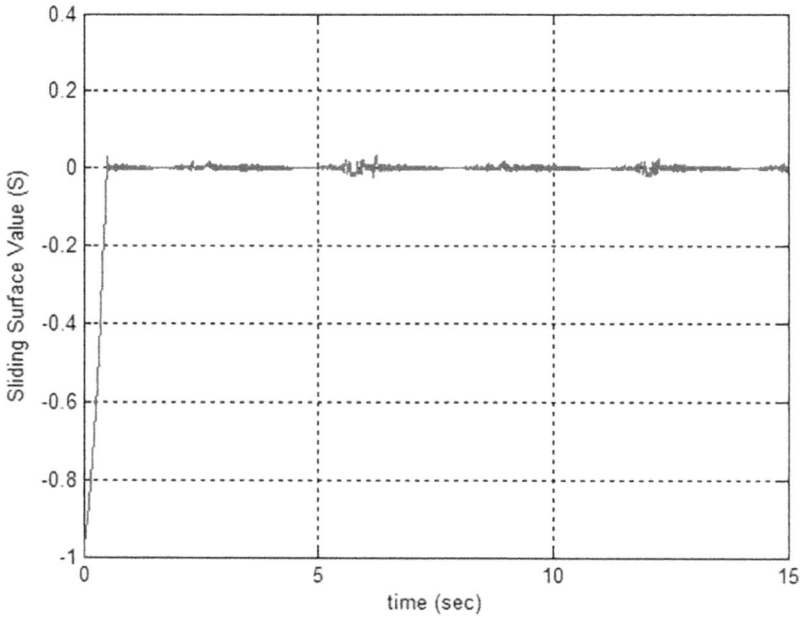

FIGURE 12.3   Sliding Surface in Example One.

12.1.2   HIGHER ORDER SLIDING MODE CONTROL WITH OPTIMAL REACHING

The sliding mode control discussed in the previous section is first order sliding mode control, in that, we strive to achieve the sliding surface $S = 0$. A higher order variation of the sliding mode control is where we wish to make the derivatives of S to approach zero as well as the S itself. In this section we discuss the higher order sliding mode control with optimal reaching [1]. Specifically, we present two different third order sliding mode control laws and compare the results. First controller we present here is the Levant third order controller given by

$$u_L = -\alpha \, Sign\left(\ddot{S} + \beta_2 \left(\left| \dot{S} \right|^3 + S^2 \right)^{\frac{1}{6}} \times Sign\left(\dot{S} + \beta_1 \left| S \right|^{\frac{2}{3}} Sign(S) \right) \right) \qquad (12.14)$$

Here, α, β_1, β_2 are known constants and $S, \dot{S}, \ddot{S}$ are the sliding surface and its derivatives. For details of the selection of the constant parameters, see [1] and the references therein. Next we present an alternate approach for third order sliding mode control that is based on the definition of manifolds and it ensures optimal reaching. The control law with optimal reaching is given by

$$u_{OR} = -\alpha \begin{cases} u_0 := 0, \quad S \in M_0 \\ u_1 := Sign\left(\ddot{S} \right), \quad S \in M_1 \setminus M_0 \\ u_2 := Sign\left(\dot{S} + \dfrac{\ddot{S}^2 u_1}{2\alpha_r} \right), \quad S \in M_2 \setminus M_1 \\ \qquad\qquad\qquad\qquad\qquad\qquad else \\ u_3 := Sign\left(\sigma(S,,\dot{S},\ddot{S}) \right), \end{cases} \qquad (12.15)$$

Where,

$$S = \left[S,,\dot{S},\ddot{S} \right]^T$$

$$\sigma\left(S,,\dot{S},\ddot{S} \right) := S + \frac{\ddot{S}^3}{3\alpha_r^2} + u2 * \left[\frac{1}{\sqrt{\alpha_r}} \left(u_2 \dot{S} + \frac{\ddot{S}^2}{2\alpha_r} \right)^{\frac{3}{2}} + \frac{\ddot{S}\ddot{S}}{\alpha_r} \right]$$

$$M_0 := \left\{ S \in \mathbb{R}^3 : S = 0, \dot{S} = 0, \ddot{S} = 0 \right\}$$

$$M_1 := \left\{ S \in \mathbb{R}^3 : S - \frac{\ddot{S}^3}{6\alpha_r^2} = 0, \dot{S} + \frac{\ddot{S}\left| \ddot{S} \right|}{2\alpha_r} = 0 \right\}$$

$$M_2 := \left\{ S \in \mathbb{R}^3 : \sigma\left(S,,\dot{S},\ddot{S} \right) = 0 \right\}$$

Also, α, α_r are known constants (details in [1]). Now we present a simulation example where we compare the results from applying the third order sliding mode control laws in (14) and (15) to a robotic car model.

Example two

Consider a robotic car model given by

$$\dot{x} = vcos(\phi), \dot{y} = vsin(\phi), \dot{\phi} = \frac{v}{l}\tan(\theta), \dot{\theta} = u$$

Here, x, y are Cartesian coordinates of the car position, θ is the direction (rotation angle) of the car, and ϕ is the rotation of the front steering wheels. Also, v, l are constant parameters where $v = 10 \, and \, l = 5$. We want to move the car in such a way that the position follow the following trajectory

$$y = 10\sin\left(\frac{x}{20}\right) + 5$$

Based on the above control objective, we define the sliding surface as

$$S = y - 10\sin\left(\frac{x}{20}\right) - 5$$

In this way, $S = 0$ achieves the control objective. Now we compute the derivatives of the sliding surface which turn out to be

$$\dot{S} = v\left(\sin(\phi) - \frac{1}{2}\cos\left(\frac{x}{20}\right)\cos(\phi)\right)$$

$$\ddot{S} = \frac{v^2}{l}\left[\tan(\theta)\left(\cos(\phi) + \frac{1}{2}\cos\left(\frac{x}{20}\right)\sin(\phi)\right) + \frac{l}{40}\sin\left(\frac{x}{20}\right)\cos^2(\phi)\right]$$

The parameters of the controllers in (14) and (15) are selected to be

$$\alpha = 20, \alpha_r = 10, \beta_1 = 1, \beta_2 = 2$$

Initial state for the simulation is set at $\left[x_0, y_0, \phi_0, \theta_0\right] = \left[0,0,0,0\right]$. Trajectory tracking results are shown in Figure 12.4. Notice that with optimal reaching, the convergence to the desirable trajectory is faster as compared to that with the Levant controller.

MATLAB Code

The MATLAB code is provided in the appendix

Exercise

Using the technique as used in example 1, plot the control signal and the sliding surface for example 2.

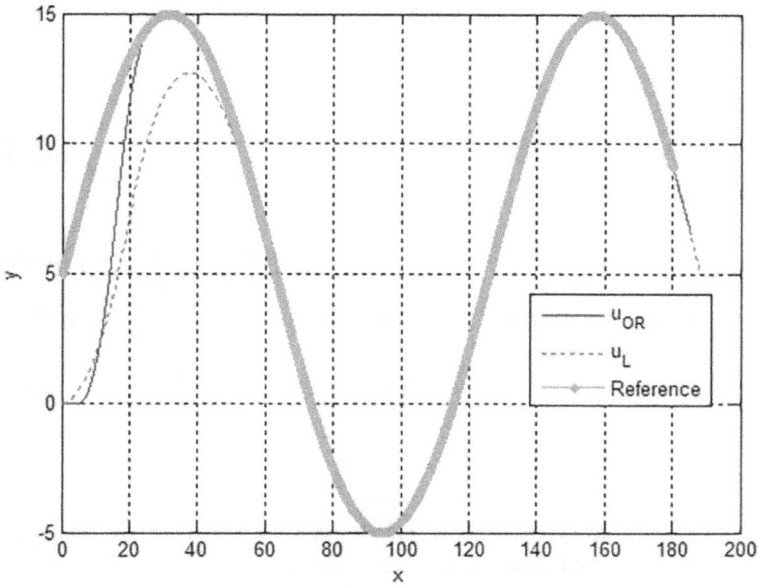

FIGURE 12.4   Trajectory Comparison with and without Optimal Reaching.

12.1.3   TERMINAL SLIDING MODE CONTROL

In this section we discuss a robust multi-input-multi-output terminal sliding mode control for robotic manipulators. There are two major differences from the conventional approach in the terminal sliding mode control discussed here. First, the sliding surface involves fractional power of the error, secondly, the robotic manipulator is controlled in such a way that it follows the behavior of a reference model (of an ideal robotic manipulator). Before we discuss the control law, we present the dynamics of n-joint robotic manipulator

$$M(q)\ddot{q} = F(q,\dot{q})\dot{q} + G(q) + u(t) \tag{12.16}$$

Here, q is $n \times 1$ vector of the angular positions of the joints, $M(q)$ is the $n \times n$ inertia matrix that is positive symmetric definite, $F(q,\dot{q})\dot{q}$ is $n \times 1$ vector of Coriolis and centrifugal torques, $G(q)$ is $n \times 1$ vector of gravitational torques.

The reference equations (ideal model) for the manipulator is defined by

$$\ddot{q}_r = Pq_r + Q\dot{q}_r + B_1 r \tag{12.17}$$

Here, $P = diag(P_i), Q = diag(Q_i), B_1 = diag(b_i), 1 \leq i \leq n$ are constant matrices which are selected in such a way that the stability of the reference model is ensured. Also, q_r is the reference joint angle vector. Note that $diag(P_i)$ is the diagonal matrix with elements $P_i, (1 \leq i \leq n)$ on the diagonal.

Now, the control law that ensures following of the reference model (17) by the manipulator with dynamics as in (16) is given by

$$
u = \begin{cases} -\dfrac{S}{a_1 \parallel S \parallel} w & \parallel S \parallel \geq \delta \\[4mm] -\dfrac{S}{a_1 \delta} w & \parallel S \parallel < \delta \end{cases}
\tag{12.18}
$$

Where a_1, δ are known constants, S is the sliding surface, and w is given by

$$
w := \parallel P q_r \parallel + \parallel Q \dot{q}_r \parallel + \parallel B_1 r \parallel + \parallel C_1 \epsilon_r \parallel + a_2 \left(\beta_1 + \beta_2 \parallel q \parallel + \beta_3 \parallel \dot{q} \parallel^2 \right)
$$

$$
\epsilon_r := diag \left(p \epsilon_1^{p-1}, \ldots, p \epsilon_n^{p-1} \right) \dot{\epsilon}
$$

$$
p = \frac{p_1}{p_2}, p_2 > p_1 \geq \frac{(p_2 + 1)}{2}
$$

Here, $\beta_1, \beta_2, \beta_3, a_2$ are known constants (bounds on uncertainty), p_2, p_1 are also constants that are chosen according to the criteria given above. Also, $\epsilon_i = q_i - q_{r,i}$ (error between i^{th} joint angle and its reference value) and ϵ is the $n \times 1$ vector of errors. Finally, $C_1 = diag(c_i), (1 \leq i \leq n)$ is a diagonal matrix of constant. The sliding surface in this case is also an $n \times 1$ vector with i^{th} element defined as

$$
S_i = c_i \epsilon_i^p + \dot{\epsilon}_i
\tag{12.19}
$$

Now we look at an example of two-joint manipulator.

Example three

Consider the 2-joint robotic manipulator with following dynamics

$$
\begin{bmatrix} \alpha_{11} & \alpha_{12} \\ \alpha_{21} & \alpha_{22} \end{bmatrix} \begin{bmatrix} \ddot{q}_1 \\ \ddot{q}_2 \end{bmatrix} = \begin{bmatrix} \beta_{12} \dot{q}_1^2 + 2 \beta_{12} \dot{q}_1 \dot{q}_2 \\ -\beta_{12} \dot{q}_2^2 \end{bmatrix} + \begin{bmatrix} \gamma_1 g \\ \gamma_2 g \end{bmatrix} + \begin{bmatrix} u_1 \\ u_2 \end{bmatrix}
$$

Where,

$$
\alpha_{11} = (m_1 + m_2) r_1^2 + m_2 r_2^2 + 2 m_2 r_1 r_2 \cos(q_2) + J_1
$$

$$
\alpha_{12} = m_2 r_2^2 + m_2 r_1 r_2 \cos(q_2)
$$

$$
\alpha_{22} = m_2 r_2^2 + J_2
$$

$$
\beta_{12} = m_2 r_1 r_2 \sin(q_2) \quad \gamma_1 = -\left((m_1 + m_2) r_1 \cos(q_2) + m_2 r_2 \cos(q_1 + q_2) \right)
$$

$$
\gamma_2 = -m_2 r_2 \cos(q_1 + q_2)
$$

The parameter values are

$$
r_1 = 1, r_2 = 0.8, J_1 = 5, J_2 = 5, m_1 = 0.5, m_2 = 1.5
$$

A reference model for the manipulator to follow is given by

$$\dot{x}_m = A_m x_m + B_m r$$

where

$$x_m = \begin{bmatrix} q_{r1} & q_{r2} & \dot{q}_{r1} & \dot{q}_{r2} \end{bmatrix}^T$$

$$A_m = \begin{bmatrix} 0 & 0 & 1 & 0 \\ 0 & 0 & 0 & 1 \\ -4 & 0 & -5 & 0 \\ 0 & -4 & 0 & -5 \end{bmatrix}, B_m = \begin{bmatrix} 0 & 0 \\ 0 & 0 \\ 1 & 0 \\ 0 & 1 \end{bmatrix}, r = \begin{bmatrix} 5 \\ 5 \end{bmatrix}$$

The parameters relevant to the controller (bounds of uncertainty) are

$$a_1 = 0.1, a_2 = 2, \beta_1 = 2, \beta_2 = 1, \beta_3 = 2$$

There shall be two sliding surface that are selected as

$$S_1 = \epsilon_1^{0.6} + \dot{\epsilon}_1$$
$$S_2 = \epsilon_2^{0.6} + \dot{\epsilon}_2$$

MATLAB simulation results with initial condition $q_{1,0} = q_{1r,0} = 0.2, q_{2,0} = q_{2r,0} = 2$, $\dot{q}_{1,0} = \dot{q}_{1r,0} = \dot{q}_{2,0} = \dot{q}_{2r,0} = 0$ are shown in Figure 12.5. Notice the absence of chattering in the control signal while reference tracking is quite good.

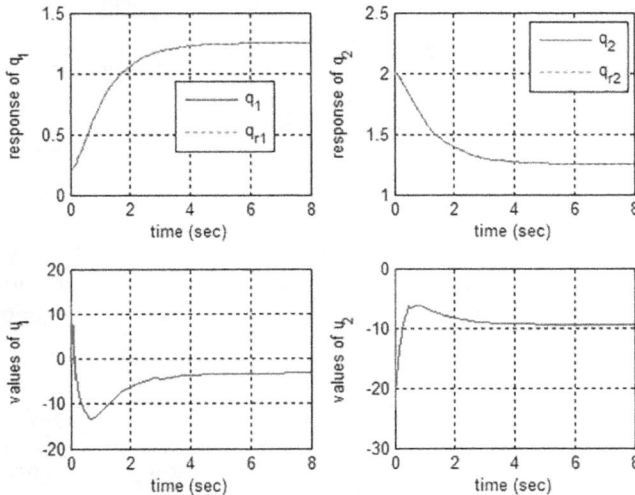

FIGURE 12.5 Reference Tracking and Control Signals for Example Three.

MATLAB Code

MATLAB code is provided in the appendix

12.1.4   SUPER-TWISTING SLIDING MODE CONTROL

Super-Twisting sliding mode control features an integral term and a term proportional to a function of the sliding surface. The control law may be presented as [3].

$$u = -K_P \sqrt{|S|} Sign(S) + v \quad \dot{v} = -K_I Sign(S) \tag{12.20}$$

Here, v is the integral state and K_P, K_I are control gains. For large enough control gains [3], the convergence of the sliding surface to zero is guaranteed.

Example 4

Consider the following state space model of an electric dynamic load simulator [3]

$$\dot{x}_1 = x_2$$

$$\dot{x}_2 = \frac{k_1 u}{k_6} - k_2 x_2 - k_3 x_1 - k_4 \ddot{x}_{1d} - k_5 \dot{x}_{1d}$$

The control objective is to make $x_1 \rightarrow x_d$ as $t \rightarrow \infty$. Also, k_2, k_3, k_4, k_5 are constants that depend upon the parameters of the load simulator. More specifically,

$$k_1 = \frac{K_G K_m}{N J_m}, k_2 = \frac{b_m}{J_m}, k_3 = \frac{K_G}{N^2 J_m}, k_4 = K_G, k_5 = \frac{K_G b_m}{J_m}, k_6 = 100$$

Where the parameters of the dynamic load simulator are selected as

$$K_m = 0.955, K_G = 85,000, J_m = 0.000697, b_m = 0.00018, N = 35$$

We select the sliding surface as

$$S = c\epsilon_1 + \epsilon_2$$

Where, $\epsilon_1 = x_1 - x_{1d}, \epsilon_2 = x_2$. We select $c = 45$. Also, for the control law in (20), the gains are selected as $K_P = 45, K_I = 990$. Simulations are performed with zero initial conditions. For learning purposes, the super twisting sliding mode control has been compared with a PID (proportional-integral) control law.

$$u = K_P \epsilon_1 + K_D \epsilon_2 + K_I \int_0^t \epsilon_1 d\tau$$

The results of the simulation of both super-twisting sliding mode control and PID control are shown in Figure 12.6 and Figure 12.7. Notice the superior performance of the sliding mode control.

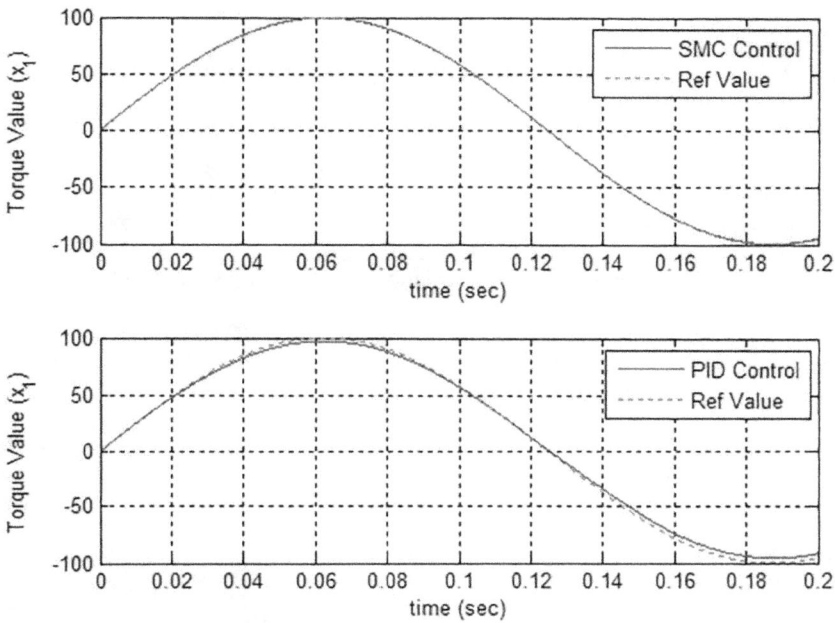

FIGURE 12.6   Reference Tracking Results for Example Four.

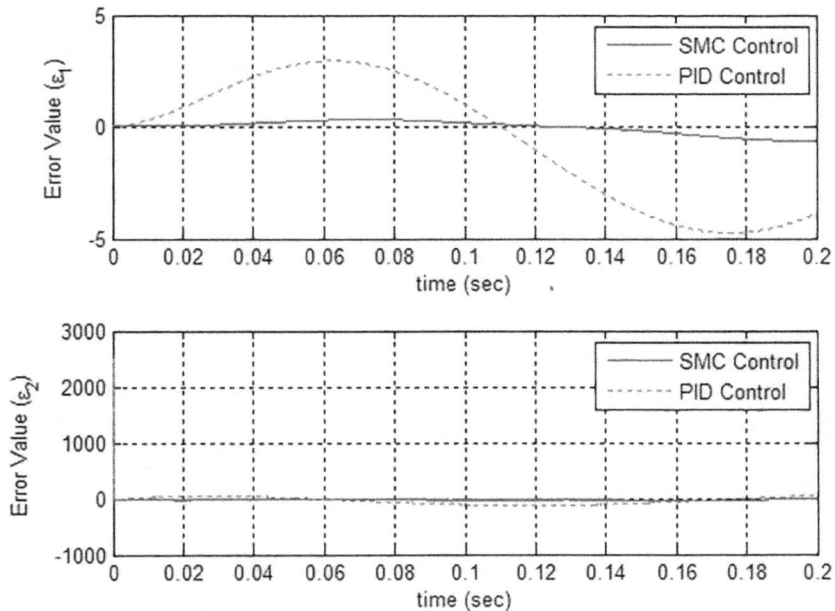

FIGURE 12.7   Error Values for Example Four.

MATLAB Code

The MATLAB code is provided in the appendix.

REFERENCES

[1] B. Bandyopadhyay, S. Janardhanan, and S. K. Spurgeon, "Advances in sliding mode control" in *Lecture Notes in Control and Information Sciences*, Springer, 440, 2013.

[2] F. Dinuzzo and A. Ferrara, "Higher order sliding mode controllers with optimal reaching," in *IEEE Transactions on Automatic Control*, vol. 54, no. 9, pp. 2126–2136, Sept. 2009.

[3] Man Zhihong, A. P. Paplinski, and H. R. Wu, "A robust MIMO terminal sliding mode control scheme for rigid robotic manipulators," in *IEEE Transactions on Automatic Control*, vol. 39, no. 12, pp. 2464–2469, Dec. 1994.

[4] M. Dai, R. Qi, and X. Cheng, "Super-twisting sliding mode control design for electric dynamic load simulator," in *2019 Chinese Control Conference (CCC)*, IEEE, 2019, pp. 3078–3083. www.ieeexplore.ieee.org

[5] J. K. Hedrick and A. Girard, *Control of Nonlinear Dynamic Systems: Theory and Applications* (Unpublished lecture notes).

[6] G. P. Incremona, M. Rubagotti, M. Tanelli, and A. Ferrara, "A general framework for switched and variable gain higher order sliding mode control," in *IEEE Transactions on Automatic Control*, vol. 66, no. 4, pp. 1718–1724, April 2021.

[7] M. B. R. Neila, and D. Tarak, "Adaptive terminal sliding mode control for rigid robotic manipulators," in *International Journal of Automation and Computing*, vol. 8, no. 2, pp. 215–220, 2011.

[8] Y. Wang, Y. Feng, X. Zhang, and J. Liang, "A new reaching law for antidisturbance sliding-mode control of PMSM speed regulation system," in *IEEE Transactions on Power Electronics*, vol. 35, no. 4, pp. 4117–4126, April 2020.

[9] V. Acary, B. Brogliato, and Y. V. Orlov, "Chattering-free digital sliding-mode control with state observer and disturbance rejection," in *IEEE Transactions on Automatic Control*, vol. 57, no. 5, pp. 1087–1101, May 2012.

[10] J. Jose and S. J. Mija, "An output feedback integral optimal sliding mode controller for magnetic levitation systems," in *2020 Fourth International Conference on Inventive Systems and Control (ICISC)*, IEEE, 2020, pp. 197–202. www.ieeexplore.ieee.org

[11] S. Y. Chen and F. J. Lin, "Robust nonsingular terminal sliding-mode control for nonlinear magnetic bearing system," in *IEEE Transactions on Control Systems Technology*, vol. 19, no. 3, pp. 636–643, May 2011.

[12] H. Bouadi, M. Bouchoucha, and M. Tadjine, "Sliding mode control based on backstepping approach for an UAV type-quadrotor," in *World Academy of Science, Engineering and Technology*, vol. 26, no. 5, pp. 22–27, 2007.

APPENDIX

INSTRUCTIONS FOR USING THE MATLAB CODE

The MATLAB code presented below is a multi-file code. For each of the example, carry out the following steps in order to reproduce the results

Step 1: Copy the code from each file of an example in a separate MATLAB file.
Step 2: Save the function file by the same name as the name of the function, e.g., for the function file in example 1, save it as "example1.m". You can

name the execution file anything (as long as the name does not contain any special character other than the underscore "_")

Step 3: Run the execution file (do not run the function file).

Note: Make sure that the function file and the execution file are stored in the same folder and the current path of the MATLAB includes that folder.

MATLAB CODE FOR EXAMPLE 1

The code is written in two files. One is the function file containing the dynamics of the closed loop system (to be solved using the differential equation solver of MATLAB). Second file is the execution file where the solution is generated by calling the function file and all the graphs are generated

File 1 (Function)

```
function dx = example1(t,x)
S = x(1)+x(2)-cos(t)-sin(t);
u = -x(1)^2-2*x(2)-0.1*exp(-t)-sin(t)+cos(t)-x(2)-
(1+abs(x(1)+x(2))+0.1/(t+1))*sign(S);
x1dot = x(2);
x2dot = x(1)^2+2*x(2)+u+0.1*exp(-t)+abs(x(1)+x(2))
*0.5+0.01/(t+1);
dx = [x1dot;x2dot];
```

File 2 (Execution)

```
clc;clear all;close all;
[t,x] = ode45(@example1,[0 15],[0 0]);
plot(t,x(:,1))
hold on
plot(t,sin(t),':r')
grid on
xlabel('time (sec)')
ylabel('State Variable Value (x_1)')
legend('x_1','x_1_d')
u = t;
S = t;
for i = 1:length(t)
    S(i) = x(i,1)+x(i,2)-cos(t(i))-sin(t(i));
    u(i) = -x(i,1)^2-2*x(i,2)-0.1*exp(-t(i))-sin(t(i))+
cos(t(i))-x(i,2)-(1+abs(x(i,1)+x(i,2))+0.1/(t(i)+1))
*sign(S(i));
end
figure(2)
```

```
plot(t,u)
grid on
xlabel('time (sec)')
ylabel('Control Input Value (u)')
figure(3)
plot(t,S)
grid on
xlabel('time (sec)')
ylabel('Sliding Surface Value (S)')
```

MATLAB CODE FOR EXAMPLE 2

File 1 (Optimal Reaching Control)

```
function dx = hosmc1(t,x)
v = 10;l = 5;a = 20;ar = 10;
s = x(2) - 10*sin(x(1)/20) - 5;
sd = v*(sin(x(3)) - 0.5*cos(x(1)/20)*cos(x(3)));
sdd = v^2/l*(tan(x(4))*(cos(x(3))+0.5*cos(x(1)/20)*sin(x
(3)))+l/40*sin(x(1)/20)*cos(x(3))^2);
u2 = sign(sd + sdd*abs(sdd)/(2*ar));
c1 = (s==0) && (sd==0) && (sdd==0);
c2 = (s-sdd^3/(6*ar^2)==0) && (sd+sdd*abs(sdd)/(2*ar)==0);
c3 = (s+sdd^3/(3*ar^2)+u2/sqrt(ar)*(u2*sd+sdd^2/(2*ar))^
(3/2)+u2*sd*sdd/ar == 0);
if t>=0.5
    if c1==1
        u = 0;
    elseif c2 == 1
        u = -a*sign(sdd);
    elseif c3 == 1
        u = -a*u2;
    else
    u = -a*sign(s+sdd^3/(3*ar^2)+u2/sqrt(ar)*(u2*sd+sdd^2/
(2*ar))^(3/2)+u2*sd*sdd/ar);
    end
else
    u = 0;
end
dx1 = v*cos(x(3));
dx2 = v*sin(x(3));
```

```
dx3 = v/l*tan(x(4));
dx4 = u;
dx = [dx1;dx2;dx3;dx4];
```

File 2 (Levant Control)

```
function dx = hosmc2(t,x)
v = 10;l = 5;a = 20;ar = 10;b1 = 1;b2 =2;
s = x(2) - 10*sin(x(1)/20) - 5;
sd = v*(sin(x(3)) - 0.5*cos(x(1)/20)*cos(x(3)));
sdd = v^2/l*(tan(x(4))*(cos(x(3))+0.5*cos(x(1)/20)*sin(x
(3)))+1/40*sin(x(1)/20)*cos(x(3))^2);
u = -a*sign(sdd+b2*(abs(sd)^3+s^2)^(1/6)*sign(sd+b1*abs
(s)^(2/3)*sign(s)));
dx1 = v*cos(x(3));
dx2 = v*sin(x(3));
dx3 = v/l*tan(x(4));
dx4 = u;
dx = [dx1;dx2;dx3;dx4];
```

File 3 (Execution)

```
clc;clear all;close all;
[t,x] = ode45(@hosmc1,[0:0.01:20],[0 0 0 0]);
[t1,x1] = ode45(@hosmc2,[0:0.05:20],[0 0 0 0]);
figure(1)
plot(x(:,1),x(:,2))
grid on
xlabel('x');ylabel('y');
hold on
plot(x1(:,1),x1(:,2),':r')
x2 = 0:0.1:180;
y2 = 10*sin(x2/20) + 5;
plot(x2,y2,'.-g')
grid on
legend('u_O_R','u_L','Reference')
```

MATLAB CODE FOR EXAMPLE 3

File 1 (Function)

```
function dx = tsmcr(t,x)
r1 = 1;r2 = 0.8;
```

```
J1 = 5; J2 = 5;
m1 = 0.5; m2 = 1.5;
g = 9.8;d = 0.01;
a1 = 0.1;a2 = 2;
b1 = 2;b2 = 1;b3 = 2;
a11 = (m1+m2)*r1^2+m2*r2^2+2*m2*r1*r2*cos(x(2))+J1;
a12 = m2*r2^2+m2*r1*r2*cos(x(2));
a22 = m2*r2^2+J2;
b12 = m2*r1*r2*sin(x(2));
g1 = -((m1+m2)*r1*cos(x(2))+m2*r2*cos(x(1)+x(2)));
g2 = -m2*r2*cos(x(1)+x(2));
M = [a11 a12;a12 a22];
MI = inv(M);
r = [5;5];
e1 = x(1)-x(5);
e2 = x(2)-x(6);
e1d = x(3)-x(7);
e2d = x(4)-x(8);
S = [e1^0.6+e1d;e2^0.6+e2d];
P = [-4 0;0 -4];Q = [-5 0;0 -5];B1 = eye(2);C1 = eye(2);
if norm(S)>=d
    er = diag([0.6*e1^-0.4,0.6*e2^-0.4])*[e1d;e2d];
%      er = [-0.6*e1^0.2;-0.6*e2^0.2];
    w = norm(P*[x(5);x(6)])+norm(Q*[x(7);x(8)])+norm(B1*
r)+norm(C1*er)+a2*(b1+b2*norm([x(1);x(2)])+b3*norm([x(3);
x(4)])^2);
    u = -S/(a1*norm(S))*w;
else
    er = [-0.6*e1^0.2;-0.6*e2^0.2];
    w = norm(P*[x(5);x(6)])+norm(Q*[x(7);x(8)])+norm(B1*
r)+norm(C1*er)+a2*(b1+b2*norm([x(1);x(2)])+b3*norm([x(3);
x(4)])^2);
    u = -S/(a1*d)*w;
end
x1dot = x(3);
x2dot = x(4);
qdot2 = MI*[(b12*x(3)^2+2*b12*x(3)*x(4)+g1*g + u(1));(-
b12*x(4)^2+g2*g + u(2))];
Am = [0 0 1 0;0 0 0 1;-4 0 -5 0;0 -4 0 -5];
```

```
Bm = [0 0;0 0;1 0;0 1];
xm = [x(5);x(6);x(7);x(8)];
xmdot = Am*xm + Bm*r;
dx = [x1dot;x2dot;qdot2;xmdot];
```

File 2 (Execution)

```
clc;clear all;close all;
[t,x] = ode23s(@tsmcr,[0,8],[0.2 2 0 0 0.2 2 0 0]);
figure(1)
subplot(221)
plot(t,x(:,1))
hold on
plot(t,x(:,5),':r')
legend('q_1','q_r_1')
grid on
xlabel('time (sec)');ylabel('response of q_1');
subplot(222)
plot(t,x(:,2))
hold on
plot(t,x(:,6),':r')
legend('q_2','q_r_2')
grid on
xlabel('time (sec)');ylabel('response of q_2');
r1 = 1;r2 = 0.8;
J1 = 5; J2 = 5;
m1 = 0.5; m2 = 1.5;
g = 9.8;d = 0.01;
a1 = 0.1;a2 = 2;
b1 = 2;b2 = 1;b3 = 2;
U = zeros(length(t),2);
for i = 1:length(t)
    a11 = (m1+m2)*r1^2+m2*r2^2+2*m2*r1*r2*cos(x(i,2)
)+J1;
    a12 = m2*r2^2+m2*r1*r2*cos(x(i,2));
    a22 = m2*r2^2+J2;
    b12 = m2*r1*r2*sin(x(i,2));
    g1 = -((m1+m2)*r1*cos(x(i,2))+m2*r2*cos(x(i,1)+
x(i,2)));
    g2 = -m2*r2*cos(x(i,1)+x(i,2));
```

```
    r = [5;5];
    e1 = x(i,1)-x(i,5);
    e2 = x(i,2)-x(i,6);
    e1d = x(i,3)-x(i,7);
    e2d = x(i,4)-x(i,8);
    S = [e1^0.6+e1d;e2^0.6+e2d];
    P = [-4 0;0 -4];Q = [-5 0;0 -5];B1 = eye(2);C1 =
eye(2);
    if norm(S)>=d
%          er = [-0.6*e1^0.2;-0.6*e2^0.2];
        er = diag([0.6*e1^-0.4,0.6*e2^-0.4])*[e1d;e2d];
        w = norm(P*[x(i,5);x(i,6)])+norm(Q*[x(i,7);x(i,8
)])+norm(B1*r)+norm(C1*er)+a2*(b1+b2*norm([x(i,1);x(i,2)
])+b3*norm([x(i,3);x(i,4)])^2);
        U(i,:) = -S/(a1*norm(S))*w;
    else
        er = [-0.6*e1^0.2;-0.6*e2^0.2];
        w = norm(P*[x(i,5);x(i,6)])+norm(Q*[x(i,7);x(i,8
)])+norm(B1*r)+norm(C1*er)+a2*(b1+b2*norm([x(i,1);x(i,2)
])+b3*norm([x(i,3);x(i,4)])^2);
        U(i,:) = -S/(a1*d)*w;
    end
end
% figure(2)
subplot(223)
plot(t,U(:,1))
grid on
xlabel('time (sec)');ylabel('values of u_1');
subplot(224)
plot(t,U(:,2))
xlabel('time (sec)');ylabel('values of u_2');
grid on
```

MATLAB CODE FOR EXAMPLE 4

File 1 (Function SMC)

```
function dx = STSMC_ELS3(t,x)
Jm = 0.000697;bm = 0.00018;N = 35;
Kg = 85000;f = 4;Km = 0.955;
```

```
thlr = 8*pi/(180*f)*sin(2*pi*f*t);
Tref = 2250*f*thlr/pi;
thlrd = 8*pi/(180*f)*2*pi*f*cos(2*pi*f*t);
thlrdd = -2*pi*f*8*pi/(180*f)*2*pi*f*sin(2*pi*f*t);
Trefd = 2250*f*thlrd/pi;
c = 45;b = 990;
e1 = Tref-x(1);e2 = Trefd-x(2);
S = e2+c*e1;
u = c*sqrt(abs(S))*sign(S)+x(3);
k1 = Kg*Km/(N*Jm);
k2 = bm/Jm;k3 = Kg/(N^2*Jm);k4 = Kg;k5 = Kg*bm/Jm;
u = k1*u*1e-2;
Phi = -k2*x(2)-k3*x(1)-k4*thlrdd-k5*thlrd;
x1dot = x(2);
x2dot = u + Phi;
x3dot = b*sign(S);
dx = [x1dot;x2dot;x3dot];
```

File 2 (Function PID)

```
function dx = STSMC_ELS2(t,x)
Jm = 0.000697;bm = 0.00018;N = 35;
Kg = 85000;Km = 0.955;
f = 4;
thlr = 8*pi/(180*f)*sin(2*pi*f*t);
Tref = 2250*f*thlr/pi;
thlrd = 8*pi/(180*f)*2*pi*f*cos(2*pi*f*t);
thlrdd = -2*pi*f*8*pi/(180*f)*2*pi*f*sin(2*pi*f*t);
Trefd = 2250*f*thlrd/pi;
e1 = Tref-x(1);e2 = Trefd-x(2);
u = 45*e1 + e2 + 990*x(3);
k1 = Kg*Km/(N*Jm);
k2 = bm/Jm;k3 = Kg/(N^2*Jm);k4 = Kg;k5 = Kg*bm/Jm;
u = k1*u*1e-2;
Phi = -k2*x(2)-k3*x(1)-k4*thlrdd-k5*thlrd;
x1dot = x(2);
x2dot = u + Phi;
x3dot = e1;
dx = [x1dot;x2dot;x3dot];
```

File 3 (Execution)

```
clc;clear all;close all;
[t1,x1] = ode45(@STSMC_ELS3,[0 0.2],[0 0 0]);
[t2,x2] = ode45(@STSMC_ELS2,[0 0.2],[0 0 0]);
f = 4;
thlr = 8*pi/(180*f)*sin(2*pi*f*t1);
Tref = 2250*f*thlr/pi;
thlrd = 8*pi/(180*f)*2*pi*f*cos(2*pi*f*t1);
thlrdd = -2*pi*f*8*pi/(180*f)*2*pi*f*sin(2*pi*f*t1);
Trefd = 2250*f*thlrd/pi;
e11 = Tref-x1(:,1);e21 = Trefd-x1(:,2);
thlr2 = 8*pi/(180*f)*sin(2*pi*f*t2);
Tref2 = 2250*f*thlr2/pi;
thlrd2 = 8*pi/(180*f)*2*pi*f*cos(2*pi*f*t2);
thlrdd2 = -2*pi*f*8*pi/(180*f)*2*pi*f*sin(2*pi*f*t2);
Trefd2 = 2250*f*thlrd2/pi;
e12 = Tref2-x2(:,1);e22 = Trefd2-x2(:,2);
figure(1)
subplot(211)
plot(t1,x1(:,1));
hold on;grid on
plot(t1,Tref,':r')
xlabel('time (sec)');ylabel('Torque Value (x_1)');
legend('SMC Control','Ref Value')
subplot(212)
plot(t2,x2(:,1));
hold on;grid on
plot(t2,Tref2,':r')
xlabel('time (sec)');ylabel('Torque Value (x_1)');
legend('PID Control','Ref Value')
figure(2)
subplot(211)
plot(t1,e11)
hold on;grid on
plot(t2,e12,':r')
xlabel('time (sec)');ylabel('Error Value (\epsilon_1)');
legend('SMC Control','PID Control')
```

```matlab
subplot(212)
plot(t1,e21)
hold on;grid on
plot(t2,e22,':r')
xlabel('time (sec)');ylabel('Error Value (\epsilon_2)');
legend('SMC Control','PID Control')
```

Index

For Product Safety Concerns and Information please contact our EU
representative GPSR@taylorandfrancis.com
Taylor & Francis Verlag GmbH, Kaufingerstraße 24, 80331 München, Germany

www.ingramcontent.com/pod-product-compliance
Lightning Source LLC
Chambersburg PA
CBHW060553220326
41598CB00024B/3087

*9 781032 116709 *